C000104277

GEOTRENDS

THE PROGRESS OF GEOLOGICAL AND GEOTECHNICAL ENGINEERING IN COLORADO AT THE CUSP OF A NEW DECADE

PROCEEDINGS OF THE 2010 BIENNIAL GEOTECHNICAL SEMINAR

November 5, 2010
Denver, Colorado

SPONSORED BY
The Geo-Institute of the American Society of Civil Engineers

Colorado Chapter of the Geo-Institute of the American Society of Civil Engineers

Rocky Mountain Section of the Association of Environmental and Engineering Geologists

Colorado Association of Geotechnical Engineers

EDITED BY
Christoph M. Goss, Ph.D., P.E.
Joseph B. Kerrigan, P.E.
Joels C. Malama, P.E.
William O. McCarron, Ph.D., P.E.
Richard L. Wiltshire, P.E.

GEO-
INSTITUTE

Published by the American Society of Civil Engineers

Cataloging-in-Publication Data on file with the Library of Congress.

American Society of Civil Engineers
1801 Alexander Bell Drive
Reston, Virginia, 20191-4400

www.pubs.asce.org

Copyright © 2011 by the American Society of Civil Engineers.
All Rights Reserved.
ISBN 978-0-7844-1144-5
Manufactured in the United States of America.

Preface

Like our dynamic earth, the practice of geotechnical engineering is always changing. Like the wind slowly carving a stone, we tweak methods developed by the soil mechanics pioneers. Like a massive rockslide, we switch from rules of thumb to detailed computer models. Like a fault, trending across the arid west, our practice seeks out weaknesses in our knowledge and technique and uplifts them to new levels of understanding. This book celebrates these GeoTrends as its papers reveal the changes in shoring and foundations, explore the role of sustainability and energy efficiency, expand our techniques for slope stability evaluation, and tie together the past and present as geophysical methods are used to locate the underground workings designed and built by our geo-ancestors.

Since 1984, the ASCE Colorado Section's Geotechnical Group, in collaboration with the Rocky Mountain Section of the Association of Environmental and Engineering Geologists and the Colorado Association of Geotechnical Engineers, has organized a biennial series of geotechnical seminars on a wide variety of themes that have been attended by as many as 270 civil/geotechnical engineers, geologists, and other geo-professionals. The geotechnical seminars have been held at area universities or hotels and have offered the opportunity for sharing ideas and experiences among Colorado's diverse geo-disciplines. Since 2004, ASCE's Geo-Institute has published the papers of these seminars in Geotechnical Practice Publications, allowing the experiences to be shared with a worldwide audience.

The GeoTrends Steering Committee convened in August 2009 and held monthly meetings to plan for the 2010 Biennial Geotechnical Seminar. The Steering Committee members included Joseph Kerrigan (Conference Chair), Dustin Bennetts, Mark Brooks, Robin Dornfest, Darin Duran, Dr. Christoph Goss, Joels Malama, Dr. Bill McCarron, Minal Parekh, Becky Roland, Keith Seaton, Jere Strickland, David Thomas, Mark Vessely Chris Wienecke, and Richard Wiltshire.

Christoph Goss, Joe Kerrigan, Joels Malama, Bill McCarron, and Richard Wiltshire

Acknowledgments

The GeoTrends Steering Committee wishes to take this opportunity to thank all of the authors and reviewers of our papers, which are herein presented as Geotechnical Practice Publication No. 6. The authors have spent many hours in preparing and finalizing their papers, which will be presented at the 2010 Biennial Geotechnical Seminar on November 5, 2010. These papers have been reviewed by a volunteer group of Denver area geo-professionals who put in their valuable time and helped make these papers even better. The Geo-Institute's Committee on Technical Publications completed its review of our GeoTrends papers in a very timely manner and their adherence to our aggressive publication schedule is greatly appreciated. We would also like to acknowledge the assistance of Donna Dickert of ASCE's Book Production Department for putting this publication together.

Contents

Thirty Years of Excavation Shoring Design and Construction Progress in Denver, Colorado

Todd L. Duncan P.E.[1], M. ASCE

[1]Branch Manager, Schnabel Foundation Company, 2950 South Jamaica Ct, Suite 107, Aurora, CO 80014; todd@schnabel.com

ABSTRACT: Schnabel Foundation Company began excavation shoring design and construction in the Denver area in May of 1979. From this first project at Writer Square to the present, hundreds of shoring projects have been completed throughout the region for commercial, residential, government, health care and transportation works. Some of the shoring systems used include driven sheeting, drilled sheeting, internal bracing, soil nailing, underpinning and micropiles.

There have been changes and improvements in many aspects of the work, including design procedures, equipment, materials, labor, and other means and methods associated with the work. There are also new challenges that have developed, such as fiber optic lines, increased utilities and directional drilling, light rail, etc. There have also been aspects of the work that have remained relatively unchanged, such as soil conditions, worker safety, etc.

This paper describes some of the history and practices in design and construction of excavation support systems, specific to the Denver, Colorado area.

INTRODUCTION

The selection and design of an excavation shoring system is comprised of many variables. First and foremost, the system must result in a reliable, safe system to protect people and property. Second, the system should be compatible with the site specific soil conditions. Third, the system should be economical, and efficient to build.

The Denver Downtown area, including Lower Downtown has seen continued growth, with occasional slow periods associated with economic conditions. For most projects some portion of the structure is frequently constructed below grade, typically for parking facilities, and the structure typically extends to the property limits. Beyond the property limits, improvement such as existing buildings, utilities, roadways, etc. may exist. Therefore, some type of shoring is used to minimize the lateral limits of the excavation and maintain uninterrupted service for adjacent property users.

1

TYPICAL SUBSURFACE CONDITIONS

The Downtown area subsurface profile is generally comprised of two soil types. The upper soils consist of alluvial sands and gravels underlain by the Denver Formation. The depth of the alluvial material varies throughout the area and ranges in depth from about six meters (20 feet) and extends to depths in excess of eighteen meters (60 feet). The material ranges in density from loose to very dense and is typically poorly graded with very minimal fine material.

The Denver Formation typically consists of weak to moderate cemented sandstone and claystone bedrock. This material extends to depths well beyond the impact of most shoring projects. The bedrock is generally impervious, however groundwater is frequently found in the bedrock and the water table in the alluvial material varies with proximity to recharge sources and is typically about one meter (3 feet) above the bedrock. Groundwater is also found in perched zones in the alluvial material.

For excavation shoring, both materials offer unique qualities and challenges for both design and construction. The alluvial material tends to cave and collapse during beam drilling and requires slurry drilling to stabilize the drill hole. If the drill depth extends to the bedrock, casing is required as the bedrock cannot be drilled efficiently under slurry.

The alluvial material has a short standup height during excavation and collapsing in common. To minimize the caving, shorter excavation lifts are done, or soil mixing may be done between soldier beams prior to the start of excavation.

SHORING SYSTEMS

The most common shoring system used in downtown Denver has been soldier beam and lagging. For excavation depths up to about four meters (13 feet), the soldier beams are typically cantilevered. Deeper excavations utilize tiebacks for lateral support. Other systems have been used, such as soil nailing, secant and tangent pile walls, and sheet piling. Constructability, economics, and other site specific requirements have generally dictated the use or lack of use of these other systems.

DESIGN METHOD

Many references and design guidelines are available for determining the lateral earth pressure for the shoring system. Most of the projects that have been designed by Schnabel Foundation Company have utilized an empirical lateral earth pressure envelope similar to those recommended by Harry Schnabel (1982). All of the projects have been completed with out failure or other excessive movement. All the monitored projects have performed well within the expected movement ranges.

Important aspects of the shoring design are the building foundation layout and the foundation construction method. The soldier beams are spaced to minimize interference with the construction of the new building foundation. In Denver, most buildings are constructed on drilled shaft foundations. The soldier beams are spaced around the drilled shaft locations to avoid drilling the caisson directly in front of the toe of the soldier beam.

Tiebacks are used to provide lateral support when an easement from adjacent property owners is obtained and when existing improvements do not prevent such installation. When tiebacks cannot be installed internal bracing may be used to provide lateral support.

FIG. 1. Internal bracing and wale, left wall. Tiebacks, right wall, Denver, CO

INSTALLATION METHODS

Driven Soldier Beams

In 1979, Writer Square, the first project that Schnabel worked on, the shoring system was installed using driven H-piles with pressure injected tiebacks. The tiebacks were connected to the soldier beams using a waler placed on the face of the soldier beam. The face of the shoring was located about two meters (6 feet) from the outside face of the building to allow adequate space for access between the formwork and the shoring and waler as the new building would be constructed using a conventional double sided form system. The void between the building and the shoring was then backfilled with soil, gravel, or other material as specified by the geotechnical engineer.

FIG. 2. Driven soldier beams with walers

Driven soldier beams can be an economical method to install soldier beams and is frequently used in many other parts of the country. In the Denver area, driven soldier beams were used until the mid 1980's, but use has largely been discontinued for the last twenty plus years.

Some of the reasons driven soldier beams are seldom used include:

1. Piles cannot be driven in the bedrock, which is regularly encountered within the required depth of the soldier beam
2. Pile driving equipment is limited in it's availability in the region as drilled shaft foundations are much more common and suitable to the local soil conditions
3. Off wall line shoring requires one to two meters (3 to 6 feet) of space beyond the building/property lines, which space typically includes utilities, buildings, or other constraints
4. Tieback connections may be more expensive
5. Shorter spans between piles, 1.75 to 2.5 meters (6 to 9 feet) on center, requiring more piles, tiebacks, and lagging connections
6. Possible vibrations from pile driving equipment may be transmitted to adjacent structures

Drilled Soldier Beams

Presently, in Denver, drilled soldier beams are considered the standard installation method for soldier beams. The soldier beam may consist of an H section, W section, or a built up section of two C or W section beams. As the hole is drilled to the required depth, casing or slurry is usually needed to facilitate drilling in Denver as the alluvial sands and gravels do not have sufficient fine material to prevent caving. If the design requires that the soldier beam extends to an elevation that is in the bedrock, the drill shaft is relatively easily advanced to the required toe depth. After the soldier beam is placed in the shaft, lean concrete (one sack cement) backfill is placed, typically end dumped, or the soil is mixed in place to form a weak soil cement mix.

Some reasons drilled soldier beams are the standard installation method:
1. Equipment and knowledge readily available
2. Reaching required design depth is easily obtained
3. More stringent installation tolerances are obtainable, very important for wall line shoring
4. Simple, straight forward tieback connections are possible
5. Greater spans between soldier beams, 2.5 to 3 meters (8 to 10 feet) on center, reducing the number of soldier beams, tiebacks, lagging connections, etc (30% reduction in quantity)
6. May be constructed directly adjacent to existing structures with minimal risk of damage

FIG. 3. Setting drilled soldier beam, Denver, CO

Tieback Installation

Tieback installation in the alluvial sands and gravels has remained relatively unchanged. Pressure grouted tiebacks have proved to be very efficient and economical. With advances in equipment for installation, grouting, and pressuring, production of anchor installation can be more than double early production rates.

To install the tiebacks in the sand, casing is drilled or driven with a sacrificial bit to the design anchor length. The tieback tendon is installed and grout is then pumped into the casing. After the casing is filled, the casing and grout are pressurized then the casing is extracted while grout pressure is maintained on the system.

As tiebacks were being installed on the first projects, little information was known about the grout soil adhesion values that may be achieved, and how much pressure was required to achieve acceptable anchor performance test results. The first anchors were installed with grout pressures five to six times the line pressure of the pump. Needless to say, some damage was inflicted on adjacent structures and corrective measures were required. As testing has continued, required grout pressures have been evaluated and typically are about 350 to 1400 kPa (50 to 200 psi) over the line pressure of the pump.

Tieback improvements in the past years are largely related to improvements in equipment that allow for more production with fewer labor hours.

Lagging

Wood lagging has been used to span between the soldier beams for many years. There are references that recommend calculations for lagging design procedures. The result of such calculations can require greatly varied lagging thickness requirements, thicknesses of 200mm (8 inches) or greater may be required. The variation in lagging design thickness is typically due to how the engineer chooses to account for the magnitude of the arching effect of the soil spanning between the driven soldier beams or the edge of the drill hole.

Many references, such as Peck (1974), ASCE GSP 74 (1997) defer to the lagging being sized by experience of the engineer and contractor that have performed acceptably for similar soil conditions. Schnabel has used 100mm (3 inch) thick lagging with soldier beams spaced at three meters (ten feet) on center, a clear span of 2.3 meters (7.5 feet) between drill holes for excavations up to 17 meters (55 feet) on numerous projects.

While the state of stress in the lagging is unknown, it is assumed that the wood is not over stressed as there is no visible excessive deformation or bowing of the lagging on the face of the shoring. Many years of experience has shown the 100mm (3 inch) lagging provides a safe, acceptable and economical product.

FIG. 4. 12.2 meter (40 foot) deep excavation with 100mm (3 inch) lagging, Colorado Springs, CO

Underpinning

When existing buildings and the proposed building will be built coincidental to the shared property line, the existing structures foundation may not extend to the same depth as the proposed structure. The existing structure will need to be supported to allow the new excavation to proceed to the lower depth.

Underpinning, either conventional hand dug or micropiling, or other method can extend the existing foundation to a new bearing elevation equal to or deeper than the new excavation. While micropiles are a newed technology, traditional hand dug underpinning is still an effective method to support the existing structure and allow the new building to efficiently utilize the entire property limits available.

Hand dung underpinning may be used when the new building design extends up to the edge of an existing adjacent building. The underpinning extends the existing foundation to a new bear elevation equal to or deeper than the new excavation. This allows an owner to utilize the entire project area.

FIG. 5. Hand dug underpinning, Denver, CO

CONCLUSIONS

Shoring systems are designed and constructed using systems that the engineer and contractor are familiar with, normally used in the area. Many parts of the country use driven systems, and tieback connections that allow some of the same advantages of drilled soldier beams. This aspect of shoring is more of a regional preference whose use is dictated by site conditions and local labor abilities.

In the Denver area, using drilled soldier beams has resulted in a decrease of about thirty percent of the soldier beams and tiebacks. The lagging area does not change, but the individual lagging connections also decreases by using the increased span of the drilled systems. Overall, this can result in a faster installation time for the project.

The drilled systems also allow the shoring system to be placed at the wall line of the new foundation, thus allowing more space at the top of the excavation and eliminating backfill of the building.

REFERENCES

ASCE (1997) "Guidelines of Engineering Practice for Braced and Tied-Back Excavations" ASCE GSP 74

Peck R., Hanson W., Thornburn T., (1959) "Foundation Engineering" John Wiley & Sons

Schnabel, H (1982) "Tiebacks in Foundation Engineering and Construction" McGraw Hill, New York

Colorado Earthquakes and Seismic Hazard

Charles S. Mueller[1]

[1]Research Geophysicist, U. S. Geological Survey, Mail Stop 966, Box 25046, Denver, CO 80225; cmueller@usgs.gov

ABSTRACT: Earthquakes are occasionally felt in Colorado, but rarely cause significant damage. About two dozen earthquakes have caused moderate, localized damage—Modified Mercalli Intensity VI or greater—in historical times. The event of November 1882, probably located north of Denver along the Front Range, is the largest known Colorado earthquake with magnitude estimates ranging from 6.2 to 6.6. The U. S. Geological Survey (USGS) makes probabilistic, national-scale seismic hazard maps that consider hazard not only from specific fault sources, but also from historical background earthquakes that may or may not be associated with known faults. Many non-tectonic earthquakes occur in Colorado, so the formulation of seismicity catalogs for hazard analysis requires special care. Compared to the 2002 edition of the USGS maps, the 2008 hazard has decreased significantly at many sites in Colorado. I illustrate these changes at four sites using some of the analysis tools that are available at the USGS website. Reasons for the differences include updates to seismicity catalogs, fault parameters, and ground-motion attenuation equations, as well as changes in background seismicity modeling.

HISTORICAL SEISMICITY

Table 1 lists about two dozen significant Colorado earthquakes compiled by Stover and Coffman (1993) using magnitude (any reported magnitude equal to or greater than 4.5) and/or intensity (Modified Mercalli Intensity, MMI, equal to or greater than VI) criteria. Stover and Coffman (1993) also provide narrative accounts of earthquake effects and damage. The largest earthquake in Colorado during historical times occurred on November 8, 1882, probably located north of Denver along the Front Range. According to Stover and Coffman (1993): "It caused minor damage in Colorado and southern Wyoming and was felt slightly in Utah and Kansas." Spence and others (1996) studied this earthquake in detail; by extrapolating intensity contours northward and eastward into sparsely settled areas (making the contours roughly symmetrical about the probable epicenter) they increased the intensity-based magnitude to 6.6 from previous estimates of 6.2 (Table 1). The next largest earthquake in the Stover and Coffman (1993) list is a magnitude-5.5 event in 1960 located near Montrose. The list could be brought up to date by adding the 2001, magnitude-4.5 earthquake near Trinidad. Another narrative history of Colorado seismicity can be found at a USGS web page: http://earthquake.usgs.gov/earthquakes/states/colorado/history.php (accessed 07 July

9

2010). A fault and epicenter map compiled by the Colorado Geological Survey is available at http://www.cdphe.state.co.us/hm/rad/rml/energyfuels/preap/ 07docs/ 07earthquake.pdf (accessed 07 July 2010). That map is also included in a pamphlet on Colorado earthquake hazards published by the Colorado Earthquake Hazards Mitigation Council (2008).

Table 1: Significant Colorado Earthquakes (Stover and Coffman, 1993)

Date	Lat. (degN)	Lon. (degE)	Mag.	MMI	Comment *
1871/10	40.5	-108.5		VI	Lily Park, Moffat County
1882/11/08	40.5	-105.5	6.2 M_{fa}	VII	North of Denver; FA=485; Spence and others (1996)
1891/12	40.5	-108.0		VI	Lily Park, Moffat County
1913/11/11	38.1	-107.7		VI	Ridgway, Ouray County; FA=14
1944/09/09	39.0	-107.5		VI	Basalt, Eagle County; FA=19
1955/08/03	38.0	-107.3		VI	Lake City, Hinsdale County; FA=5
1960/10/11	38.3	-107.6	5.5 m_b	VI	Montrose; FA=39
1962/02/05	38.2	-107.6	4.7 M_L	V	Montrose
1962/12/04	39.8	-104.7	3.2 M_L	VI	Denver**; FA=12
1962/12/05	39.9	-104.6	3.8 M_L	VI	Denver**; FA=16
1965/02/16	39.9	-105.0	3.0 M_L	VI	Denver**; FA=1
1965/09/14	39.9	-104.6	3.6 M_L	VI	Denver**; FA=3
1965/09/29	39.8	-105.1	3.5 M_L	VI	Denver**; FA=4
1965/11/21	39.8	-104.8	3.8 M_L	VI	Denver**; FA=7
1966/10/03	37.4	-104.1	4.6 M_L	VI	Trinidad; FA=45
1966/11/14	39.9	-104.7	3.5 M_L	VI	Denver**; FA=4
1967/04/10	39.94	-104.75	4.3 Mn	VI	Denver**; FA=16
1967/04/27	39.91	-104.77	3.8 M_L	VI	Denver**; FA=4
1967/08/09	39.9	-104.7	4.9 m_b	VII	Denver**; FA=50
1967/11/27	39.87	-104.88	4.6 m_b	VI	Denver**; FA=56
1979/01/06	38.96	-105.16	2.9 M_L	VI	Cripple Creek, Teller County; FA=11
1981/04/02	39.91	-104.95	3.8 M_L	VI	Denver; FA=6

* FA is felt area in 1,000 km^2
** Denver earthquakes during 1962–67 thought to be related to fluid disposal at Rocky Mountain Arsenal

The USGS makes probabilistic, national-scale seismic hazard maps for use in building-codes, emergency-planning, land-use planning, insurance, and other practical applications (Frankel and others, 1996, 2000, 2002; Petersen and others, 2008). The 48-state maps are updated about every six years, and new seismicity catalogs are prepared for each update. Two separate catalogs with somewhat different properties are constructed: one for the tectonically active western United States (WUS), and one for a less-active region combining the central and eastern United States (CEUS) craton, the Rocky Mountains, and the Colorado Plateau. We currently lack Rocky

Mountain- and Colorado Plateau-specific ground-motion relations for hazard modeling. The association of these two regions with the CEUS is based on assumed similarity of ground-motion attenuation properties. The combined CEUS – Rocky Mountain – Colorado Plateau region is referred to simply as CEUS hereinafter. The WUS/CEUS boundary passes through Colorado, somewhat complicating the construction of a single seismicity catalog for the state. The WUS catalog covers a region where seismicity is relatively plentiful, and where the sources of seismic hazard are well defined at the magnitude-4 level (in other words, extending the catalog to smaller magnitudes does not illuminate significant new sources of hazard). On the other hand, in the sparsely seismic CEUS all earthquakes are needed; here we extend the catalog down to magnitude 3 (where we still have some confidence in its completeness). By tradition, different magnitude scales have generally been used in seismic-observatory and hazard-analysis practice in the two regions: moment magnitude (M_w) is currently preferred in the WUS, and Lg-wave (or Nuttli) magnitude (m_{bLg}) in the CEUS. The CEUS catalog extends back to 1700, while the WUS catalog only extends back to 1850 (with catalog completeness differences accounted for in the hazard analysis). Thus, the WUS catalog is M_w-based, 1850-to-present, and M_w-4-and-greater, while the CEUS catalog is m_{bLg}-based, 1700-to-present, and m_{bLg}-3-and-greater. The two catalogs are mapped in Figure 1. The map area in Figure 1 is all CEUS except for the thumb-shaped zone of WUS that extends northward from New Mexico into the San Luis basin / Rio Grande rift area of southern Colorado.

Because non-tectonic seismic events (usually "man-made" earthquakes) generally follow different recurrence laws than natural seismicity, it is important to identify and separate them when formulating seismic-hazard models. Figure 1 shows four prominent clusters of non-tectonic seismicity in Colorado. Earthquakes occurred near Rangely following commencement of fluid injection for secondary hydrocarbon recovery in 1957 (Raleigh and others, 1976). Earthquakes occurred at the Rocky Mountain Arsenal (RMA) near Denver following the commencement of waste-fluid disposal in a 12,000-ft well in 1962 (Healy and others, 1968) (Table 1). Seismicity in the Paradox Valley is apparently related to injection of saline ground water that commenced in the 1980s (Ake and others, 2005). Earthquakes near Paonia are probably related to coal mining (V. Matthews, personal communication). These events are removed from the catalog for the hazard analysis. An earthquake at RMA in 1981 is included as a tectonic event because it is judged to have occurred sufficiently long after the cessation of fluid-disposal activities. A few other mining-related events (blasts, collapses, etc.) are identified from other sources and removed. Three nuclear explosions, part of the Atomic Energy Commission's Operation Plowshare, are also removed: 1967 Gasbuggy (Farmington, NM), 1969 Rulison (Grand Valley, CO), and 1973 Rio Blanco (Rifle, CO).

We make the traditional assumption that earthquakes are statistically independent in the seismic hazard analysis, so the catalogs are declustered to remove obvious foreshocks and aftershocks (for details see Petersen and others, 2008). The final catalogs (declustered, with non-tectonic seismicity removed) are mapped in Figure 2. The catalogs (and supporting information) can be found at http://earthquake.usgs.gov /hazards/products/conterminous/2008/catalogs (accessed 07 July 2010).

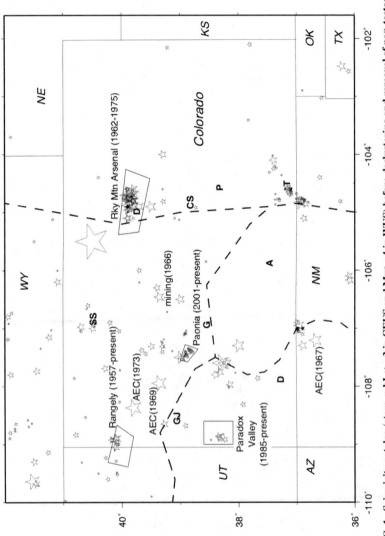

FIG. 1. Seismicity catalog (stars, mbLg≥3 in CEUS and Mw≥4 in WUS) before declustering and removal of non-tectonic earthquakes. Heavy dashed lines separate tectonic regions (see text). Polygons enclose clusters of non-tectonic earthquakes; other non-tectonic events are represented by italic and bold letters, respectively. States and Colorado cities are labeled. States and Colorado cities are represented by italic and bold letters, respectively.

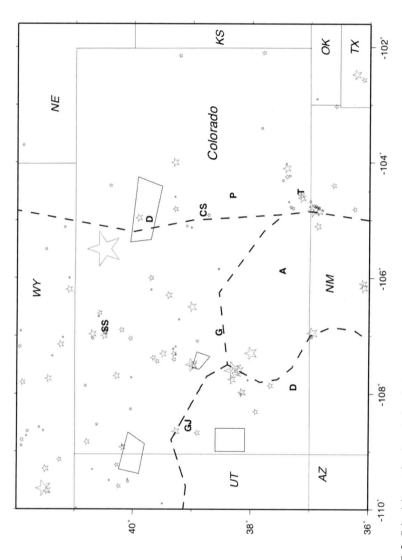

FIG. 2. Seismicity catalog (stars) after declustering and removal of non-tectonic earthquakes. See Figure 1 caption for map details.

FAULTS

Where recurrence or slip-rate data are available from paleoseismic or geologic studies, specific fault sources can be incorporated into the USGS seismic-hazard model. The best information on recurrence comes from analysis of dateable deposits found in research trenches or natural fault exposures, or in earthquake-related deposits like liquefaction sand-blows. Fault slip-rate information comes from measured offsets of dateable geological features like stream channels or river terraces or deposits. These kinds of field studies are generally labor-intensive and expensive, and data are available for only a few of the many faults in and near Colorado. Faults that are included in the USGS hazard model as specific sources are mapped in Figure 3; most have a normal-slip faulting style and relatively small slip rates. The USGS on-line Quaternary fault and fold database can be found at http://earthquake.usgs.gov/hazards/qfaults (accessed 07 July 2010). Details about the subset of faults that are actually modeled in the USGS national seismic-hazard maps are presented at http://gldims.cr.usgs.gov/webapps/cfusion/sites/hazfaults_search/ hf_search_main.cfm?hazmap=2007 (accessed 07 July 2010).

SEISMIC HAZARD

The USGS national seismic-hazard maps and details of the probabilistic methodology are presented by Frankel and others (1996, 2000, 2002) and Petersen and others (2008). Maps, reports, documentation, related hazard products, and web-based analysis tools (including results for Alaska, Hawaii, Puerto Rico, other regions, and special studies) are presented at http://earthquake.usgs.gov/hazards (accessed 07 July 2010).

For each site on a grid, ground-motion exceedance rates (hazard curves) from all modeled seismic sources are computed using published ground-motion-prediction (attenuation) relations. In most cases we apply a uniform firm-rock site condition, corresponding to a near-surface shear-wave velocity (Vs30) of 760 m/s (2,490 ft/s). Sources include specific faults and seismic rate grids derived from historical seismicity; magnitude ranges are adjusted in the model to prevent over- or under-counting of hazard contributions. For suspected active faults that lack recurrence or slip-rate data, we make the traditional assumption that their hazard is modeled by the historical seismicity. Hazard maps are generated for several structural periods and probability levels, and engineering-design maps are derived from the hazard maps (Leyendecker and others, 2000).

Updates of the national seismic-hazard maps (1996, 2002, 2008; tied to building-code update cycles) are always based on the best current geoscience. Compared to the 2002 edition of the maps, the 2008 hazard has decreased significantly (10 to 20 percent or more) at many sites in Colorado for all structural periods and probability levels. Changes at four sites are illustrated here using some of the tools that are available at the USGS website http://earthquake.usgs.gov/hazards, "Online Seismic Analysis Tools" (accessed 07 July 2010). Table 2 shows probabilistic ground motions for 10 percent probability of exceedance in 50 years (abbreviated 10px50 here, corresponding to 0.0021 annualized exceedance rate): peak ground acceleration

(PGA), 0.2-second spectral response for 5 percent damping (0.2sSA), and 1.0-second spectral response for 5 percent damping (1.0sSA). Table 3 shows corresponding numbers for 2-percent probability of exceedance in 50 years (02px50, 0.0004 annualized exceedance rate). Note that the structural-design maps used in the current NEHRP Provisions and International Building Code are based on modified versions of the 0.2sSA and 1.0sSA, 2-percent-in-50-year hazard maps (Leyendecker and others, 2000). (Data sources for Tables 2 and 3: http://eqint.cr.usgs.gov/deaggint/2002 for 2002 and http://eqint.cr.usgs.gov/deaggint/2008 for 2008, both accessed 07 July 2010.)

Table 2. Probabilistic Ground-motions (*g*, 10px50) from 2002 and 2008 at Four Sites

		Denver	Alamosa	Grand Junction	Trinidad
PGA	2002	0.035	0.059	0.050	0.041
	2008	0.031	0.057	0.042	0.055
0.2sSA (5% damping)	2002	0.079	0.135	0.107	0.088
	2008	0.067	0.132	0.089	0.115
1.0sSA (5% damping)	2002	0.022	0.040	0.026	0.026
	2008	0.020	0.035	0.023	0.028

Table 3. Probabilistic Ground-motions (*g*, 02px50) from 2002 and 2008 at Four Sites

		Denver	Alamosa	Grand Junction	Trinidad
PGA	2002	0.110	0.146	0.150	0.127
	2008	0.093	0.148	0.122	0.194
0.2sSA (5% damping)	2002	0.219	0.341	0.286	0.245
	2008	0.184	0.352	0.235	0.350
1.0sSA (5% damping)	2002	0.057	0.110	0.067	0.068
	2008	0.050	0.093	0.059	0.074

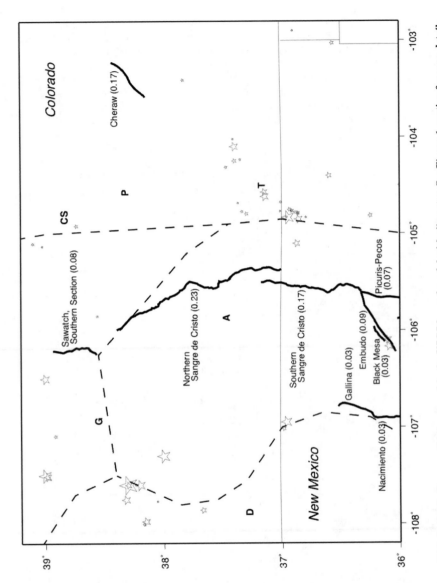

FIG. 3. Faults (heavy solid lines) used in the USGS hazard model with slip rates. See Figure 1 caption for map details.

There are many reasons for the differences, including updates to seismicity catalogs, fault parameters, and ground-motion attenuation models (predicted ground motions were generally smaller in 2008, significantly smaller for some structural periods), as well as changes in modeling details. The assumed dip on most normal faults in the Intermountain West region was decreased from 60 degrees to 50 +/- 10 degrees in the 2008 update; for a given assigned slip rate, this geometrical effect can increase modeled seismicity rates by 20 percent or more (K. Haller and S. Harmsen, Appendix J in Petersen and others, 2008). In the 2008 update we accounted explicitly for magnitude uncertainty when computing seismic-activity rates from earthquake catalogs. Because of the magnitude-frequency distribution of earthquakes—there are many more small earthquakes than large ones, accounting for magnitude uncertainty always has the effect of reducing modeled seismicity rates for a given catalog. For example, in the presence of statistical uncertainty, a reported magnitude-5.0 earthquake is much more likely to be a true magnitude-4.9 earthquake than a 5.1. For most sites in Colorado, the combined effect of magnitude uncertainty and attenuation-model updates accounts for most of the decrease in hazard. The hazard at Alamosa decreases less than elsewhere (and even increases for some structural periods and probability levels), probably due to its proximity to the Northern Sangre de Christo fault (Figure 3), a west-dipping normal fault (see above). Routine seismicity-catalog updates can sometimes bring interesting local consequences. The hazard at Trinidad increased significantly in 2008, due to a local sequence of earthquakes that commenced in 2001. These are assumed to be tectonic earthquakes, and because the sequence was long-lived, enough events survived the catalog-declustering process to strongly increase the local hazard.

CONCLUSIONS

About two dozen earthquakes have caused moderate, localized damage in Colorado in historical times. The largest known event, with estimated magnitude 6.2 to 6.6, occurred north of Denver along the Front Range in 1882. Preparation of seismicity catalogs for seismic-hazard analysis is complicated by non-tectonic earthquakes in Colorado. In addition to background seismicity, several Colorado faults are included as specific sources in the USGS national seismic hazard maps. Compared to 2002, the hazard is smaller at most sites in Colorado in the 2008 maps. Most of the decreases can be attributed to updating ground-motion attenuation models and accounting for magnitude uncertainty in background seismicity. Increases at a few sites are due to local seismicity or changes in fault geometry modeling.

ACKNOWLEDGEMENTS

Reviews by Kathy Haller, Melanie Walling, and Paul Thenhaus are gratefully acknowledged.

REFERENCES

Ake, J., Mahrer, K., O'Connell, D., and Block, L. (2005). "Deep injection and closely monitored induced seismicity at Paradox Valley, Colorado", *Bull. Seism. Soc. Am.*, **95** (2), 664–683.

Colorado Earthquake Hazards Mitigation Council (2008). Colorado Earthquake Hazards, 1-page, fold-out pamphlet, with map. (also http://geosurvey.state.co.us/Portals/0/Earthquake_Map_2008.pdf, accessed 07 July 2010).

Frankel, A.D., Mueller, C.S., Barnhard, T.P., Perkins, D.M., Leyendecker, E.V., Dickman, N., Hanson, S.L., and Hopper, M.G. (1996). National seismic-hazard maps—Documentation June 1996, *U.S. Geological Survey Open-File Report 96-0532,* 110 pages.

Frankel, A.D., Mueller, C.S., Barnhard, T.P., Leyendecker, E.V., Wesson, R.L., Harmsen, S.C., Klein, F.W., Perkins, D.M., Dickman, N.C., Hanson, S.L., and Hopper, M.G. (2000). USGS national seismic-hazard maps, *Earthquake Spectra,* **16** (1), 1–19.

Frankel, A.D., Petersen, M.D., Mueller, C.S., Haller, K.M., Wheeler, R.L., Leyendecker, E.V., Wesson, R.L., Harmsen, S.C., Cramer, C.H., Perkins, D.M., and Rukstales, K.S. (2002). Documentation for the 2002 update of the National Seismic Hazard Maps, *USGS Open-File Report 2002–420,* 39 pages.

Healy, J.H., Rubey, W.W., Griggs, D.T. and Raleigh, C.B. (1968). "The Denver Earthquakes", *Science* **161** (3848), 1301–1310.

Leyendecker, E.V., Hunt, R.J., Frankel, A.D., and Rukstales, K.S. (2000). Development of maximum considered earthquake ground motion maps, *Earthquake Spectra,* **16** (1), 21–40.

Petersen, M.D., Frankel, A.D., Harmsen, S.C., Mueller, C.S., Haller, K.M., Wheeler, R.L., Wesson, R.L., Zeng, Y., Boyd, O.S., Perkins, D.M., Luco, N., Field, E.H., Wills, C.J., and Rukstales, K.S. (2008). Documentation for the 2008 Update of the United States National Seismic Hazard Maps, *U. S. Geological Survey Open-File Report 2008-1128,* 60 pages plus appendixes.

Raleigh, C.B., Healy, J.H., and Bredehoeft, J.D. (1976). "An experiment in earthquake control at Rangely, Colorado", *Science*, **191** (4233), 1230–1237.

Spence, W., Langer, C.J., and Choy, G.L. (1996). "Rare, large earthquakes at the Laramide deformation front—Colorado (1882) and Wyoming (1984)", *Bull. Seism. Soc. Am.*, **86** (6), 1804–1819.

Stover, C.W., and J.L. Coffman (1993). *Seismicity of the United States, 1568–1989 (Revised)*, U. S. Geological Survey Professional Paper 1527, 418 p.

Geotechnical Trends for Sustainable and Constructible Solutions

Eric D. Bernhardt[1], P.E. and Robin Dornfest[2], C.P.G., P.G.

[1]Project Manager, CTL Thompson, Inc., 351 Linden Street, Unit 140, Fort Collins, Colorado 80524; ebernhardt@ctlthompson.com
[2]Geotechnical Department Manager – Senior Engineering Geologist, CTL Thompson, Inc., 351 Linden Street, Unit 140, Fort Collins, Colorado 80524; rdornfest@ctlthompson.com

ABSTRACT: Geotechnical consultants can significantly contribute to reducing impacts on our environment by providing sustainable and constructible alternatives if they are incorporated early in the project. Residential and commercial developments are "going green" by incorporating sustainable design into projects and geotechnical consultants are providing alternatives to help achieve these goals. Recent requirements and the desire to attain Leadership in Energy and Environmental Design (LEED) accreditation for projects have driven all aspects of construction into new and challenging directions.

Geotechnical consultants have been providing sustainable alternatives for many years. Due to the increased buzz around environmentally Low Impact Development (LID) or "green" building, there has been a renewed demand for these types of design solutions. To accomplish these goals, geotechnical consultants need to identify and fully understand the project needs as well as what is important to the project team and owner. Often, geotechnical consultants do not present solutions that may have significantly less impact on the environment to the project design team due to incorrect assumptions of the project goals. If sustainability is a high priority to the project team and owner, the geotechnical engineer may be able to provide creative solutions which may differ significantly from common practice but meet project goals.

Recent practices in development have called for geotechnical consultants to provide a wide range of sustainable design alternatives for all aspects of the project including: stormwater infiltration, use of recycled materials, limiting transportation of materials to sites, reuse of materials on-site, and geothermal heating and cooling. Geotechnical consultants that are thoroughly integrated with the design team can significantly contribute to overall sustainable design and construction practices for the project. Conducting "sustainability due diligence investigations," can help other design team members understand the feasibility of a variety of sustainable alternatives.

INTRODUCTION

Geotechnical Engineering and Sustainability

Geotechnical consultants are in a position to significantly influence project direction during the "big picture" stages of decision making. Successful contribution from the geotechnical consultant requires a change in mindset to include: interdisciplinary approach; new types of investigations, data collection, modeling, and analysis; better understanding of project goals; broadly educated workforce; and entrenchment into the design team at all phases of design and construction. This paper discusses when and how geotechnical engineering consultants can provide sustainable and constructible alternatives that will significantly reduce impacts on our environment.

History

The field of geotechnical engineering has been filled with a variety of creative practitioners who have evolved over time to provide useful and effective solutions and sustainable solutions represent the next stage of evolution. Geotechnical engineering has matured and the problems facing these consultants have changed significantly over the past 50 years. For instance, as expansive soils and bedrock in Colorado, Texas and many other places in the world have caused distress to structures and roadways, geotechnical engineers have developed unique techniques to mitigate adverse affects using engineering judgment and innovation. Geotechnical engineering consultants have also been providing sustainable solutions for hundreds of years. Only recently have these solutions been given a name and classified as "sustainable" or "green" alternatives which have renewed their demand.

Municipal and federal agencies, as well as other governing organizations have implemented increased and varied requirements for residential and commercial developments. Among those is the desire for a project to be built to Leadership in Energy and Environmental Design (LEED) standards whether it is slated for LEED certification or not. This desire has triggered an evolution in design teams where all levels of development contributors, including geotechnical consultants, need to incorporate sustainable solutions into their respective design element.

What is LEED?

As defined by the U.S. Green Building Council (USGBC), "LEED is an internationally recognized green building certification system, providing third-party verification that a building or community was designed and built using strategies aimed at improving performance across all the metrics that matter most: energy savings, water efficiency, CO_2 emissions reduction, improved indoor environmental quality, and stewardship of resource and sensitivity to their impacts" (USGBC).

SUSTAINABLE EXAMPLES

Geotechnical consultants have been providing green and sustainable alternatives for many years. The intent of this paper is not to list and discuss all of the geotechnical related sustainable alternatives and the associated benefits and disadvantages of each alternative in this paper as new sustainable ideas, concepts and technology are being developed quickly. Instead, we have chosen to discuss some of the current sustainable examples and how these alternatives can help the design team improve sustainability.

Geogrid. The geotechnical consultant can provide alternative pavement sections and stabilization recommendations to include geogrid which can significantly reduce the amount of imported natural resources including asphaltic concrete, aggregate base course, and/or crushed rock required for typical pavement sections. This will result in a of reduced number of truck trips to the site for delivery of construction materials, reduced excavation depths, and reduced thicknesses for pavement sections which reduces the overall impact to the environment.

Reclaimed Asphalt Pavement (RAP). The geotechnical consultant can recommend use of recycled materials in pavement construction. Over the past several decades, a common sustainable solution for asphaltic pavements has included use of reclaimed asphalt pavement or RAP. Recent research by the Colorado Department of Transportation (CDOT) has shown that RAP has displayed similar performance properties to traditional crushed rock used as aggregate base course (Locander, 2009). CDOT has allowed use of RAP as an equivalent material for use as aggregate base course in some applications. Existing asphalt pavements can also be recycled and blended into the underlying subgrade to enhance the performance and improve the structural support for the pavement system to reduce the thickness of the new pavement section. This reduces the need for import material and the costs and impacts associated with the removal of the old asphalt.

Geoexchange Systems. The geotechnical consultant can provide thermal conductivity and thermal diffusivity testing along with evaluation of geothermal gradients to assist the project team with evaluation of geoexchange systems. Subsurface investigation by the geotechnical consultant can provide the design team with recommendations on whether a vertical system or a horizontal system may be more appropriate for the project site. Geoexchange is the most energy-efficient, environmentally clean, and cost-effective space conditioning system available, according to the Environmental Protection Agency (EPA) (www.geoexchange.org). These systems can be incorporated and evaluated by the geotechnical consultant to substantially reduce energy use.

Sub-Excavation. As an alternative to importing non-expansive or higher strength fill materials to a project site, a geotechnical engineering consultant can provide alternative recommendations to include sub-excavation or over-excavation of existing on-site soils and/or bedrock to be reused at the site as moisture conditioned and properly compacted fill. Reusing site-derived materials reduces or even eliminates the amount of delivery trips to the site. In addition, it greatly reduces the use of nonrenewable natural resources that would be used for import fill.

Infiltration. By evaluating the infiltration characteristics of the ground below a project site, the geotechnical consultant can assist the design team with determining if infiltration concepts can be used to capture surface water runoff and allow the runoff to infiltrate into the ground. Innovative placement of an infiltration system can increase the usable portion of the site, improve water quality, recharge ground water, and reduce the amount of runoff flowing into streams and rivers. Examples of infiltration systems include pervious concrete, permeable interlocking concrete pavers, and infiltration galleries.

Recycled Materials. The geotechnical consultant can recommend alternative materials to be considered by the design team that can provide similar, or better, performance. For instance, recycled concrete may be readily available very near the project site which can be considered as an alternative to aggregate base course or retaining wall backfill. Shredded tires may be used as an alternative to coarse aggregate required for septic tank absorption systems or drain fields. Use of these types of recycled materials can significantly reduce use of nonrenewable resources and number of delivery trips to a project site.

Fly Ash. The geotechnical consultant can provide recommendations for use of fly ash in concrete as an alternative to conserve portland and divert fly ash from landfills. Coal fired power plants produce more than half of the electricity we consume in the United States every day (www.flyash.com). In addition to electricity, these plants produce fly ash as a byproduct, which is becoming a vital and sustainable ingredient for improving the performance of a wide range of concrete products. Fly ash is a pozzolan that, when mixed with lime (calcium hydroxide), combines to form cementitious compounds. Concrete containing fly ash can be stronger, more durable, and more resistant to chemical attack. This product can also reduce the amount of energy necessary to generate portland cement and reduce greenhouse gas emissions. In addition, fly ash can be mixed with subgrade soils to improve soil strength.

Deep Foundation Alternatives. Deep foundation alternatives can be considered by the geotechnical consultant to reduce impacts to projects located in environmentally sensitive areas including wetlands, contaminated sites and historical areas. As an example, helical piles offer a versatile and efficient alternative to conventional deep foundations or anchors in a wide variety of applications (Perko 2009). In general, manufacturing a helical pile foundation consumes much less material than conventional deep foundation alternatives and require fewer delivery trips to the project site than a conventional deep foundation system. Helical piles made with recycled oil filed drilling pipes have also been implemented into project plans. Stone columns, rammed aggregate piers and other deep foundation alternatives can also utilize on-site materials or recycled materials.

Selective Site Grading. The geotechnical consultant can assist the project team with identifying different materials and associated soil properties during site development. For projects with a significant amount of cut and fill, soils identified as having poor characteristics can be selectively removed or stockpiled during site grading activities and placed as waste berms, landscaped areas or sound barriers. Alternatively, soils identified as having potentially good properties can also be stockpiled and screened for bedding material, drain rock or structure backfill. Processing these on-site materials normally imported to the site can reduce delivery trips and save nonrenewable natural resources.

GEOTECHNICAL CONSULTANT'S ROLE

Each sustainable solution has associated constraints. Bringing a geotechnical consultant on board during the initial phases of project development to assist with "big picture" decisions can enhance their overall contribution to the project's sustainability goals. By integrating with design teams during these initial stages of design and planning, geotechnical consultants are able to help the team take a proactive rather than reactive approach to implementing sustainability into their projects when dealing with subsurface conditions at the site.

Challenges

One challenge the geotechnical community has faced when incorporating sustainable solutions into their recommendations is the associated risk and uncertainty. Much of the geotechnical engineering community relies upon historical performance of geotechnical concepts because of the inherent uncertainty of soil and bedrock properties. Traditionally, geotechnical consultants prefer to witness their innovative concepts succeed for a long period of time before including them with their array of regularly recommended concepts accepted as standard practice. Many of these sustainable solutions are unproven, have a very short track record, and are often not considered standard practice. Balancing the benefits of a sustainable alternative against the uncertain risk associated with that particular alternative is a challenge. Either way, the geotechnical consultant should educate the project team about risks and benefits associated with each recommended alternative.

Another challenge for the geotechnical consultant involves adequately informing and educating the design team and owner about how much a sustainable alternative may vary in cost, schedule impacts, or life cycle considerations from a more commonly implemented alternative or standard of practice. Many times maintenance and future improvements can be substantially more expensive and difficult when using sustainable alternatives. The geotechnical consultant should educate the project team about how the "life" of the material or solution may provide more or less green benefit for the entire life of the improvement. For example, if a sustainable pavement type has a longer potential pavement life than a less sustainable alternative, the benefit may be realized with a reduction in maintenance and/or extended life of the pavement system that would postpone the more aggressive long-term repairs and/or reconstruction. For instance, repairing a utility below a pervious pavement section involves a more difficult excavation and repair. A certified pervious pavement contractor may need to be involved with each breach through the pavement to ensure the pervious characteristics of the pavement remain intact. Similar challenges may be experienced when excavations extend through a geogrid placed below a pavement section.

It is often difficult for geotechnical engineering consultants to embed their contributions and presence into the project team very early in the project schedule and continue their involvement through completion. Typically, the geotechnical consultant is engaged in the project during initial design phases to provide design and construction criteria to other design team members. Once these criteria are determined and shared with the other design team members, the geotechnical consultant's involvement with the project generally ends. It is important that the geotechnical consultant understands the project goals. By communicating regularly with a project team, each team member can be educated as to how the geotechnical engineering consultant can contribute to all stages of the project design and construction. The challenge for the geotechnical consultant is to educate the architectural and engineering communities to understand the importance of including the geotechnical consultant in each stage of the project.

All of the project consultants, including geotechnical, may be faced with incorporating preferred sustainable solutions into the project design when the specific project site may not be suitable to support the proposed sustainable solution. Pressure to consider or implement unproven, ill-fitting, or risky alternatives into a situation can be challenging to a geotechnical consultant. Again, early participation from geotechnical consultants can help the project team avoid moving too far into the conceptual design with an unrealistic or inappropriate sustainable concept.

The geotechnical consultant can become a valuable project team member by familiarizing himself/herself with recent developments in sustainable technology and solutions. The push by the green movement/culture to utilize green concepts on projects has spurred research for performance and utilization of recycled products and materials that impact the environment less than more conventional alternatives. These ideas and sustainable solutions are coming from all aspects of society including universities, contractors, architects, engineers, municipalities, and more. It is very challenging for the geotechnical community to become experts in sustainable research and advancement in technology and new ideas. Keeping up with and understanding the available solutions is challenging, however, necessary to help the project team evaluate additional alternatives and understand their impact on the proposed project.

Geotechnical Due Diligence Considering Sustainability

As mentioned previously, in order for the geotechnical consultants to successfully contribute to the project goals, they need to stay involved during all aspects of the design and planning as well as during construction. Other design consultants should be interested in how the subsurface conditions below the project site will affect the proposed sustainable solutions as well as what sustainable solutions may be available that have not been considered prior to the geotechnical consultant's involvement.

Conducting "sustainability due diligence investigations," can help other design team members understand the feasibility of a variety of sustainable alternatives. For instance, during this type of investigation, a geotechnical consultant may identify expansive soils underlying a site that may preclude the use of stormwater infiltration through pervious pavements or other systems without significant risk of structure or pavement distress caused by swelling of expansive soils/bedrock. Or, a "sustainability due diligence investigation," may identify subsurface conditions that support use of a vertical geoexchange system with much more efficiency than a horizontal geoexchange system. Preliminary "sustainability due diligence investigations', can significantly contribute to the project direction and guide "big picture" sustainable concepts.

The scope of services that a geotechnical consultant would provide for a "sustainable due diligence investigation," might be significantly different than typically provided during a more conventional due diligence investigation of a site. Typical questions answered by a geotechnical consultant during a conventional due diligence or geologic and preliminary geotechnical investigation or evaluation of a project site would likely include the following:

1. Did you identify any geologic hazards or geotechnical constraints that would preclude development of the site?
2. What type of foundations and floor systems are appropriate for the subsurface conditions identified on the site?
3. Will the amount of water-soluble sulfates in the soils and/or bedrock below the site present risk for sulfate attack on concrete?
4. What pavement type and thickness should be expected for this site?

A different group of questions would be answered during a "sustainable due diligence investigation," and would likely include the following:

1. Are the subsurface conditions appropriate for pervious pavements?
2. What is the infiltration rate of the soils and/or bedrock below the site?
3. What is the thermal conductivity of the ground and is a geothermal system feasible?
4. Are there subsurface conditions and local codes that would not allow use of recycled rubber tire shreds as the drain material for the drain field?
5. Are there recycled materials available that can be used as structural fill or to improve the pavement subgrade?

CONCLUSIONS

When brought in at an early stage, geotechnical consultants can be a valuable contributor to the design team when considering sustainable and constructible solutions. Although geotechnical consultants have been providing sustainable solutions for many years, it is important that geotechnical consultants still rely upon traditional solutions while incorporating a new mindset to present and consider more innovative ideas. There are many challenges associated with this ever-changing green movement/culture and geotechnical consultants need to entrench themselves into the process. An effective way for the geotechnical consultant to successfully contribute to incorporating sustainability into a project is to perform a "sustainability due diligence investigation" for the project during preliminary and conceptual design phases to help identify what, if any, sustainable solutions are appropriate and what alternatives may not be appropriate given the subsurface conditions.

REFERENCES

Locander, R. (2009). "Analysis of Using Reclaimed Asphalt Pavement (RAP) as a Base Course Material." *Colorado Department of Transportation Research Branch,* Report No. CDOT-2009-5.

Perko, H.A. (2009). "Helical Piles – A Practical Guide to Design and Installation." John Wiley & Sons, Inc., Hoboken, NJ.

USGBC, U.S. Green Building Council, n.d., Web., 1 May 2010.

Engineering Performance of Thermo-Active Foundations

John S. McCartney, Ph.D., P.E.[1], M. ASCE, Joshua E. Rosenberg[2], S.M. ASCE, and Axaule Sultanova[3], S.M. ASCE

[1]Assistant Professor and Barry Faculty Fellow, University of Colorado at Boulder. Dept. of Civil, Env. and Arch. Eng.. UCB 428 Boulder, CO 80309. john.mccartney@colorado.edu.
[2]Graduate Research Assistant, University of Colorado at Boulder. Dept. of Civil, Env. and Arch. Eng.. UCB 428 Boulder, CO 80309. joshuaerosenberg@gmail.com.
[3]Undergraduate Research Assistant, University of Colorado at Boulder. Dept. of Civil, Env. and Arch. Eng.. UCB 428 Boulder, CO 80309. axaule.sultanova@colorado.edu.

ABSTRACT: Incorporation of heat exchangers into drilled shaft foundations is a novel approach to improve the energy efficiency of building heat pump systems and provide necessary structural support using the same construction materials. An efficiency evaluation indicates that heat exchange systems such as thermo-active foundations are particularly suited to the climate in Colorado. This study focuses on the changes in the ultimate axial capacity of thermo-active foundations that can be expected after cycles of heating and cooling through a review of the literature from Europe. To compliment the data in the literature, a preliminary series of centrifuge-scale model tests were performed on thermo-active foundations in compacted silt. Foundations that were heated to 60 °C after application of a building load indicate an increase in axial capacity of 30% compared to a baseline foundation tested at 25 °C. Foundations that were heated to 60 °C after application of a building load, then cooled back to 25 °C showed an increase in capacity of only 20% above the baseline case.

INTRODUCTION

Commercial and residential buildings consume 71% of the electricity generated in the U.S. and 53% of its natural gas (EIA 2008). Buildings consume approximately 39% of the primary energy in the U.S., of which heating and building systems consume 20% of this fraction. Because of this large energy use, buildings can be attributed to generating 43% of the U.S. carbon emissions (EIA 2008). Development and characterization of new technologies to reduce building energy consumption are important goals for the United States from both environmental and economic perspectives. Ground source heat pumps (GSHPs) have been used in Colorado and other regions of the United States for many years as they require less energy to heat and cool buildings than conventional air-source heat pumps. This is because GSHPs exchange heat with the subsurface soil and rock, which has a relatively steady temperature throughout the year compared with that of the outside air. Although the subsurface temperatures vary with geologic setting throughout the U.S., the average temperature of the ground below a depth of 1.3 meters is approximately 10 to 15 °C year-round (Omer 2008). Although the U.S. was a pioneer in the development of modern GSHP systems, their relatively high installation costs have led to less significant rates of implementation than other energy efficiency technologies (Hughes 2008). Their installation involves insertion of polyethylene loops into boreholes filled with sand-bentonite outside of the building footprint. Not only does this require significant labor, but requires open space near the building to install the boreholes, trenching beneath the frost depth to connect the heat exchangers, and potential horizontal directional drilling to connect the borehole field to a heat pump system within the building.

To counter this high installation cost, drilled shaft foundations are an alternate pathway to the subsurface that can be exploited for use as a GSHP while at the same time providing necessary structural support for the building. These systems may not only be more cost-effective to install because they use construction materials for multiple purposes, but they may be more efficient heat exchangers due to the high thermal conductivity and heat capacity of concrete compared to soil. Thermo-active foundations have been successfully implemented in buildings in Europe (Brandl 2006; Laloui et al. 2006; Adam and Markiewicz 2009), Japan (Ooka et al. 2007), and the UK (Bourne-Webb et al. 2009; Wood et al. 2009), they have not been widely implemented in the United States. This is partially due to questions about soil-structure interaction that may occur due to thermal expansion and contraction of the foundation and soil. If not properly designed, thermal soil-structure interaction may lead to differential foundation movements or changes in axial capacity.

To compliment observations from the literature, this study includes an evaluation of the axial capacity of scale-model thermo-active foundations in compacted silt tested in the geotechnical centrifuge at the University of Colorado at Boulder. The goal of these load tests is to understand the effects of temperature fluctuations on the deformation and structural capacity of thermo-active foundations under representative building loads. This evaluation involves performing load tests on the scale-model foundations which have experienced different temperature fluctuations.

GROUND-SOURCE HEAT EXCHANGE SYSTEMS

GSHPs were first developed by the Austrian mining engineer Peter Ritter von Ritinger in 1855, and have since undergone a steady improvement in installation techniques and heat pump infrastructure (Brandl 2006). GSHPs exchange heat between the ground and a heat pump system within the building by circulating an antifreeze fluid (i.e., propylene glycol) through the polyethylene tube in the borehole. A typical heat pump cycle in heating mode is shown in Figure 1(a). A compressor is used to decrease the volume of the refrigerant, increasing its temperature. This heated refrigerant is circulated through an air handler to transfer heat to the building or to warm water inside a water tank (potentially connected with a closed loop of coiled pipes embedded in the floor slab for ambient heating). The refrigerant then goes through an expansion valve, which leads to a decrease in its temperature. The refrigerant then absorbs heat from the antifreeze by passing through a heat exchange coil, shown in Figures 1(b) and 1(c), after which the process is repeated. This process can be reversed to supply cooling.

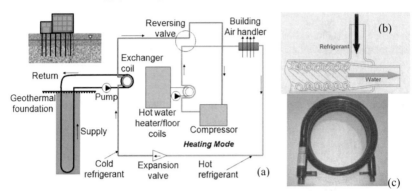

FIG. 1: Ground-Source Heat Pump components: (a) Heat pump loop; (b) Heat exchanger coil schematic; (c) Heat exchanger coil picture

The usage cost associated with heat pump systems is associated with the frequency at which a compressor must be powered to expand or contract the refrigerant. In air source heat pumps, this rate typically depends on the exterior air temperature, which can be highly variable over time. The goal of a GSHP system is to reduce the frequency at which the compressor must be powered on average throughout the year by evening out the extreme seasonal variations in temperature. Conventional borehole GSHPs, typically consisting of 5 cm diameter polyethylene tubing bent into a "U"-shape, are usually installed into 100-mm diameter boreholes to depths up to 60 meters below the ground surface (McCartney et al. 2010). While conventional heat exchange systems employ convective air flow to exchange heat into the atmosphere, GSHPs employ only conduction to exchange heat into the ground. This may potentially require a significant length of heat exchanger pipe, depending on the soil type and ground water level. The total piping requirements range from 60 to 180 meters per cooling ton (approximately 11.5 kW) depending on local soil types, groundwater levels, and temperature profiles (Omer 2008). The use of propylene glycol as a heat exchange loop permits heat to be extracted from the ground even under sub-freezing conditions. The required flow rate through the primary heat exchanger is typically between 1.5 and 3.0 gallons per minute per system cooling ton (0.027 and 0.054 L/s-kW) (Omer 2008). The typical temperatures in the heat exchanger fluid for GSHPs range from -5 to 50 °C (Brandl 2006).

A recent study by Krarti and Studer (2009) has shown that GSHPs reduce electricity peak demand, energy use, and CO_2 emissions by 10-30% in a typical Colorado home compared to conventional heating and cooling systems. The approach of Krarti and Studer (2009) was used to compare the energy usage for GHSP and conventional HVAC systems for a simple detached residence in different parts of the U.S. Specifically, the eQuest3D interface for the DOE-2 software was used to

compare the energy usage of vertical loop GSHP systems with conventional
furnace/HVAC system for different climates. For the vertical loop analyses, the
number and length of the GSHP heat exchanger loop lengths were selected for each
climate based on available design guidelines (Kavanaugh et al. 1997), and ranged
from 4 to 10 boreholes with lengths of 50 to 60 m. For simplicity, the house
incorporated insulation in all walls, the ceiling, and floor in all locations, and the same
soil thermal properties were used in all of the locations throughout the country. The
results of the eQuest3D analyses are shown in Figure 2(a) for the heating energy
consumed, in Figure 2(b) for the cooling energy consumed, and Figure 2(c) for the
total annual energy consumed. The GSHP systems provided total energy savings
between 10 and 45%, consistent with the results from Krarti and Studer (2009).

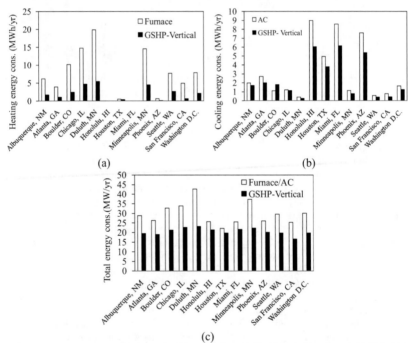

(a)   (b)

(c)

**FIG. 2: Performance of furnace/AC, vertical well GSHP, and horizontal slinky
GSHP heating/cooling systems: (a) Heating energy consumed; (b) Cooling
energy consumed; (c) Total energy consumed**

Despite the clear annual energy savings by GSHPs, the GSHP system provided
different performance with respect to heating and cooling, with a better improvement
in energy savings for heating-dominated climates due to the greater energy
consumption of furnace systems compared to AC systems. Accordingly, the regions of
the US in which GSHPs did not provide a significant decrease in energy consumption
are those in cooling-dominated locations like Honolulu, Houston, and Miami. Cold

climates such as Duluth and Minneapolis may also have to install a back-up heating system to respond to rapid changes in temperature because GSHPs often show an inertial response to changes in temperature (Kavanaugh et al. 1997). In summary, the results indicate that GSHP systems may work best for locations such as Boulder, Albuquerque, Chicago, and Washington D.C. because the GSHP system can sufficiently perform the job of a furnace and an AC system. This means that Colorado is particularly suited for GSHP implementation.

THERMO-ACTIVE FOUNDATIONS

The term thermo-active foundation is used in this study as it is inclusive of all types of building foundations used as heat exchangers (deep, shallow, driven, drilled, etc.). Outside of the U.S., thermo-active foundations are typically referred to as "Energy Piles" (Amis et al. 2009). This name arose because the first thermo-active foundations consisted of driven precast concrete piles with embedded heat exchange loops. The majority of thermo-active foundations involve heat exchangers attached to the reinforcement cage in drilled or augured foundations. Brandl (2006) reported that there are currently over 25,000 thermo-active foundations in Austria, with installations dating as early as the 1980's. Over the past five years, the installation of thermo-active foundations has grown exponentially in the UK (Amis et al. 2008). There were approximately eight times more thermo-active foundations installed in 2008 than in 2005. The reason for this rise in production is mainly driven by the code for sustainable buildings that requires the construction of zero-carbon buildings by 2019 (Bourne-Webb et al. 2009). Similar targets are being set across the globe suggesting a continued increase in production of such systems throughout the world.

The majority of thermo-active foundations involve polyethylene heat exchangers ("U"-tubes) attached to the reinforcement cage for drilled foundations as shown in Figure 3(a). Drilled shaft foundations are constructed by drilling a large diameter hole (0.4 m to 1.5 m) into the ground, lowering the steel reinforcement cage with the heat exchangers into the hole as shown in Figure 3(b), and backfilling the hole with concrete. Heat exchange loops are typically tied together with header pipes at the ground surface as shown in Figure 3(c) and connected to the heat pump system for a building. Accordingly, a distinct advantage of thermo-active foundations over conventional borehole GSHP systems is that land is not needed outside of the building footprint for heat exchange, and the heat pump infrastructure and connections are within the building footprint. This can be a major advantage in metropolitan areas like Denver and Colorado Springs.

To maximize the number of loops per foundation, the typical diameter of thermo-active foundations typically ranges from 0.5 to 4 meters (Ooka et al. 2007). If the system is coupled into the foundation plan, the cost of boring becomes essentially zero, further reducing the capital cost. The capital cost of the heat exchange component of a thermo-active foundation is that of the HDPE tubing and the labor cost of connecting the tubing to the steel reinforcement of the drilled shaft foundation. Additional costs come from quality assurance testing (i.e., pressurized leak tests) needed for thermo-active foundations to avoid punctures in heat exchange loops during assembly or installation (Brandl 2006).

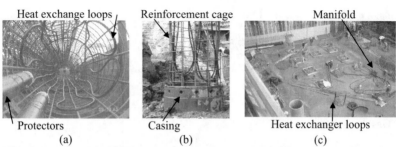

Heat exchange loops   Reinforcement cage   Manifold

Protectors   Casing   Heat exchanger loops
(a)   (b)   (c)

**FIG. 3: Installation of heat exchangers in thermo-active foundations (after
Ebnother 2008): (a) Heat exchange loops on reinforcement cage; (b)
Inserting reinforcement into cased hole; (c) Connection of heat exchangers
after installation**

THERMAL RESPONSE OF THERMO-ACTIVE FOUNDATIONS

A few full-scale thermo-active foundation projects have provided some energy
usage data (Ooka et al. 2007; Adam and Markiewicz 2009; Wood et al. 2009), proving
the feasibility of this approach to provide sustainable heat output and long-term
reductions in heating and cooling costs. The thermal performance of GSHPs is
typically defined using the Coefficient of Performance (COP), which is equal to the
thermal energy delivered by the system divided by the electricity input to operate the
system. A typical COP value for air-source heat pumps (ASHPs) is 1-3 (Brandl 2006),
although this varies with climate as indicated by the analysis in Figure 2. GSHPs
typically have a COP greater than 3, although this number may be lower for particular
thermo-active applications.

Wood et al. (2009) constructed a test plot consisting of 21 thermo-active
foundations which were 10 meters deep. A heat rejection setup was devised to
simulate the heat load of a two-storey, modern residential building with a ground floor
area of 72 m^2. Testing of the thermo-active foundations was performed over a heating
season, with two different heating loads. The heating load and COP for the system is
shown in Figure 4(a). The COP of the system was relatively consistent throughout the
test, and equal to approximately 3.75. Ooka et al. (2007) compared the COP of a
thermo-active foundation and an air-source heat pump over the period spanning
cooling and heating seasons, as shown in Figure 4(b). The COP of the thermo-active
foundation was found to be twice as high as the ASHP during the cooling season
(early times), but decreases to 1.5 times greater during the heating season (late times).
Adam and Markiewicz (2009) evaluated the performance of several different
thermally-active geotechnical systems, including foundations, tunnel linings, sewers,
and diaphragm walls. For a thermo-active foundation used to support a cut-and cover
tunnel, they observed a COP of approximately 2 for the period of several years, as
shown in Figure 4(c). This system is exposed to the air in the tunnel, which may
explain why its COP is not as high as the others measured in the literature.

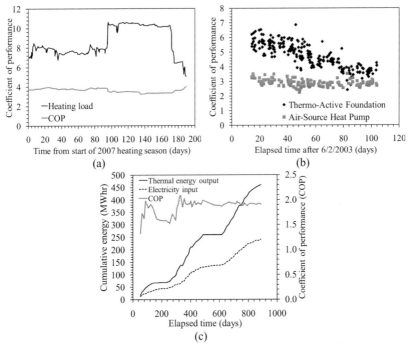

FIG. 4: Thermal performance of thermo-active foundations: (a) Coefficient of
performance during different magnitudes of heating load (after Wood et
al. 2009); (b) Coefficient of performance over a summer and winter (after
Ooka et al. 2007); (c) Coefficient of performance of a support foundation
for a cut and cover tunnel (Adam and Markiewicz 2009)

Modeling of the thermal response of thermo-active foundations is challenging due
to the need to understand the building heating and cooling load for a particular
building design and climate. For example, a poorly insulated building in a cold climate
has little chance of having successful thermal performance with a GSHP or ASHP
system. The state-of-the-art in thermal modeling is to use simplified quasi-analytical
solutions to quantify the length of heat exchanger needed for a given building load
(Kavanaugh 1997), which are incorporated into available GSHP design software
(GLHEPro, EQuest, ECA, etc.). However, the quasi-analytical solutions available
consider only a few conventional GSHP heat exchanger configurations (spacing and
diameters), and do not consider conditions representative of thermo-active
foundations. Accordingly, design tools are needed for the thermal analysis of thermo-
active foundations to ensure sufficient thermal mass to heat and cool the building in a
sustainable fashion. Advanced heat exchange models such as EnergyPlus have been
used to evaluate building slab heat losses (Krarti 1995), and are being evaluated for
use in modeling heat exchange in thermo-active foundations.

The groundwater table may have implications on the required length of heat exchangers in deep foundations as the degree of saturation can affect the thermal conductivity of soils (Abu-Hamdeh and Reeder 2000). The soil mineralogy may also play an equally important role to the degree of saturation. Tarnawski et al. (2009) observed that the quartz content has a significant effect on soil thermal conductivity. Because of these issues, site-specific thermal capacity measurements are recommended (Shonder and Beck 1997).

MECHANICAL RESPONSE OF THERMOACTIVE FOUNDATIONS

In addition to the heat exchange design for thermo-active foundations, geotechnical design is also required. Geotechnical design requires consideration of the complex interaction between temperature change and induced stresses and strains in the foundation, which may affect building performance. Specifically, contraction or expansion of the foundation during cooling or heating may lead to changes in foundation side friction (and ultimate foundation capacity) or mechanical distress in the concrete. Further, extreme conditions issues such as frost heave and subsequent settlement upon melting may occur if heat exchanger fluid temperatures are reduced below freezing for extended periods of time. Although freezing conditions in the heat exchanger can be avoided by programming of the control system for the heat pump, malfunctions may occur. This study is concerned primarily with the changes in ultimate capacity of the foundation during heating and cooling.

Deformations may occur in thermo-active foundations due to the phenomenon of thermo-elastic expansion, in which thermal strain ε_t occurs during a change in temperature proportionally to a coefficient of thermal expansion ($\varepsilon_t = \alpha_T \Delta T$) The coefficient of thermal expansion α_T of concrete can be as high as 14.5 x 10^{-6} m/m °C, while that of the steel used as reinforcement is 11.9 x 10^{-6} m/m °C (Choi and Chen 2005). These values indicate that thermal expansion of reinforced concrete will be similar to these two materials since they are approximately compatible. Choi and Chen (2005) and Bourne-Webb et al. (2009) have quantified the thermal strain in reinforced concrete, with their results shown in Figure 5. Although it appears that Bourne-Webb et al. (2009) measured a higher coefficient of thermal expansion for a foundation, the response was affected by the measurement technique (fiber-optic cable). Correction of the data led to a coefficient of thermal expansion of 8.5 x 10^{-6} m/m °C, consistent with the data of Choi and Chen (2005). The amount of thermal expansion or contraction for a thermo-active foundation will depend on soil-structure interaction, as the surrounding soil may provide a confining effect on the foundation.

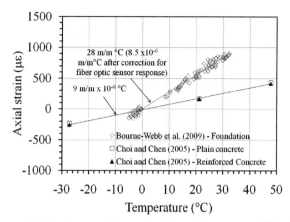

FIG. 5: Thermal strain in concrete as a function of temperature (Choi and Chen 2005; Bourne-Webb et al. 2009)

The mechanisms of thermo-mechanical effects on thermo-active foundations can be evaluated by assessing data presented by Bourne-Webb et al. (2009), who performed a series of thermal and mechanical loading tests on a full-scale foundation in England, and Laloui and Nuth (2006), who performed a series of thermal and mechanical loading tests on a full-scale foundation in Switzerland. The foundation tested by Bourne-Webb et al. (2009) was a 0.56 m diameter drilled shaft with a depth of 22.5 m, containing three polyethylene heat exchange loops. The lower 18.5 m of the foundation is in London clay with the rest of the foundation in cohesionless and fill material. The foundation tested by Laloui and Nuth (2006) was a 25.8 m-long drilled shaft having a diameter of 0.88 m. The upper 12 m of the foundation was in alluvial soils, while the lower part of the foundation was in glacial moraine material.

The coupled thermo-mechanical loads in thermo-active foundation produce unique stress and strain profiles, shown schematically in Figures 6(a) and 6(b) (after Bourne-Webb et al. 2009). When a foundation is loaded under a mechanical load the largest stresses are seen at the top and diminish with depth, as shown in the left-hand schematic in Figures 6(a) and 6(b). This loading profile is representative of the case in which the end bearing is not fully mobilized and when there are no residual stresses in the foundation from installation. If the pile is heated, it will expand volumetrically. Although the temperature distribution in thermo-active foundations during heating is complex because heat is shed along the length of the heat exchanger tube, it can be assumed that the foundation changes in temperature uniformly for simplicity. Bourne-Webb et al. (2009) assumed that the foundation expands about its mid-point in this situation, as shown in the central schematic in Figure 6(a). This assumption implies that the soil shear resistance is constant with depth, and that heat is not shed linearly along the length of the heat exchanger into the soil. Nonetheless, the thermal expansion will result in an increase in compressive stress throughout the foundation due to the axial expansion, and an increase in side friction due to the radial expansion. For this simple explanation, the coupled response produces a uniform stress in the upper portion of the pier and the total stresses could be twice those seen from the mechanical load alone as shown the right-hand schematic in Figure 6(a) (Bourne-Webb et al. 2009; Laloui et al. 2006). Although the side friction will increase uniformly with depth during heating, the direction of the side shear force will be opposite on either side of the mid-point of the foundation since it is assumed to expand from the mid-point. As the foundation is cooled, it will tend to contract volumetrically. Because the mechanical load diminishes toward the bottom, tensile forces could occur in the foundation if the cooling load is significant (Bourne-Webb et al. 2009). More importantly, contraction during cooling will lead to a reduction in radial stresses which could possibly lead to a decrease in side friction. Although cycles of heating and cooling may lead to a cumulative decrease in side friction if the soil does not rebound after heating, this has not been thoroughly investigated.

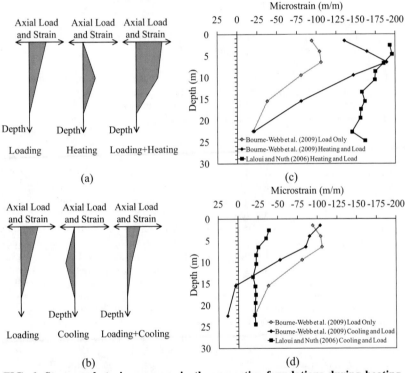

FIG. 6: Stress and strain response in thermo-active foundations during heating and cooling (after Bourne-Webb *et al.* 2009 and Laloui and Nuth 2006): (a) Schematic of stress/strain changes during heating; (b) Schematic of stress/strain changes during cooling; (c) Measurements of strains during heating; (b) Measurements of strains during cooling.

The schematics of the strain profiles hypothesized for thermo-active foundations can be assessed by comparing the results from the load tests available in the literature. Bourne-Webb et al. (2009) loaded their foundation to 1200 kN, cooled it to -6 °C, and then heated it to 40 °C. Laloui and Nuth (2006) loaded their foundation to 2140 kN, increased the temperature by 21 °C above the natural ground temperature, then cooled it to 3 °C above the natural ground temperature. The strain distributions in the foundations tested by Bourne-Webb et al. (2009) and Laloui and Nuth (2006) after initial loading (data was only available from Bourne-Webb et al. 2009) and heating are shown in Figure 6(c). These strains were measured using fiber optic cables in the case of Bourne-Webb et al. (2009) and with strain gauges in the case of Laloui and Nuth (2006). The initial strain value at the bottom of the foundation measured by Bourne-Webb et al. (2009) indicates that there was a slight mobilization of end bearing. Similarly, the strain distributions after loading and cooling are shown in Figure 6(d). A small tensile stress was noted in the bottom of the foundation tested by Bourne-Webb et al. (2009) because the end bearing had not been fully mobilized. Overall, the observations from the field after loading then heating or cooling are consistent with the schematic strain distributions in Figures 6(a) and 6(b).

These effects could lead to heave or settlements of the foundation butt, and could potentially create down-drag on the foundation. For a foundation that was not loaded axially, Laloui et al. (2006) observed a butt heave of nearly 4 mm during an increase in temperature of 21 °C over the period of 1 day. The foundation did not return to its original elevation upon cooling, but maintained an upward displacement of approximately 1 mm. The amount of movement and stresses in the foundation depends on the end-restraints of the foundation by the lower bearing stratum and the building load (Bourne-Webb et al. 2009). Although the movements are minor, Laloui et al. (2006) indicated that the increase in temperature may have led to a plastic response in the clay. The soil was observed to partially recover deformations after cycles of heating and cooling, causing permanent foundation movement (Laloui et al. 2006).

The most significant risk of thermo-active foundations is in the possibility for differential movements as asymmetric thermal expansion or contraction could lead to the generation of bending moments and differential movement. Should heat exchange loops fail or clog in a given foundation, the foundation will cease to change in temperature. Significant differential expansion or contraction could occur should the heat exchange loops in a particular foundation fail next to a fully-functional foundation (Laloui et al. 2006). Boennec (2009) indicates that the current design practice in Europe is to assume that 10% of the heat exchange tubes can be expected to fail during the lifetime of a foundation. Differential displacements may also occur near the outer boundary of the building, where internal temperatures may be different from outer temperatures. These effects can be considered by limiting the range of temperature fluctuations, and possibly changing reinforcement patterns.

To consider the implications of thermal-induced movements in the foundation, engineers in Switzerland double the design factor of safety for ultimate capacity for thermo-active foundations from that used for conventional foundation design (Boënnec 2009). Bourne-Webb et al. (2009) reported that a design safety factor of 3.5 for ultimate capacity was used in the design of the thermo-active foundation system for Lambeth College in the UK. The justification for such conservatism in safety factors is being investigated in recent research studies throughout the world, as it may effectively require twice as many foundations to support the same building load.

In addition to structural concerns, practical issues such as matching the number, depth, and spacing of ground loops with the required number, depth, and spacing of deep foundations must also be considered in the design of thermo-active foundation systems. If more ground loops than deep foundations are required, then an auxiliary conventional GSHP system may need to be used outside of the building footprint. McCartney et al. (2010) found that the thermal and structural requirements of a typical building can be attained with the same number of foundations, assuming AASTHO LRFD foundation resistance factors are used in foundation structural design.

MODELING OF FOUNDATION PERFORMANCE

The results of Bourne-Webb et al. (2009) and Laloui and Nuth (2006) helped identify the mechanisms governing the structural performance of thermo-active foundations. However, their tests required significant cost and time to perform. The field results are sensitive to the foundation installation process and the soil profile at each site. In order to build upon their experience, a series of scale-model foundations were evaluated in the geotechnical centrifuge at the University of Colorado at Boulder [Figure 7]. Scale-model testing in the centrifuge permits parametric evaluation of the variables affecting the structural response of thermo-active foundations under controlled conditions. The setup used to evaluate the performance of thermo-active foundations is shown in Figure 8. A cylindrical, insulated aluminum tank with an inside diameter of 0.8 m and height of 0.7 m was used to evaluate the performance of three thermo-active foundations having a diameter of 76.2 mm and length of 381 mm. Four foundations were placed within the container at a spacing of 3 diameters at a given time. This spacing was found to lead to minimal interference between foundations with respect to thermal and mechanical loading, while maintaining the same soil conditions for each of the foundations. The concrete foundations were cast outside the centrifuge using welded wire mesh as the reinforcement cage and an aluminum pipe as the heat exchanger. A heat pump outside the centrifuge was used to circulate fluid through the heat exchanger to heat and cool the foundation. The foundations were tested in silt, which was compacted around the foundations in lifts to a dry unit weight of 17.2 kN/m^3.

FIG. 7: Geotechnical centrifuge at the University of Colorado at Boulder

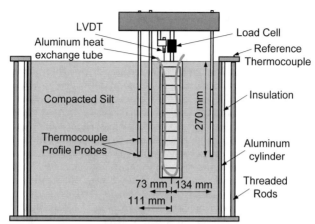

FIG. 8: Centrifuge-scale testing setup for thermo-active foundations

The results from three scale-model foundations performed as part of a preliminary investigation into centrifuge-modeling of thermo-active foundations are presented in this paper. The foundations were tested at a g-level of 24. At this g-level, they represent prototype foundations with a length and diameter of 9.1 m and 1.8 m, respectively. Each of the foundations was tested individually in subsequent order, after the soil and foundations had returned to ambient conditions. Temperature profiles indicate that temperature did not extend to the soil in the vicinity of the other foundations. The load settlement curves (in prototype scale) for the three foundations are shown in Figure 9. The load-settlement curve for a baseline foundation (Test 1) was obtained by applying a constant displacement rate of 0.08 mm/min to the butt of the foundation and measuring the load. The impact of heating and cooling was evaluated for two other foundations (Tests 2 and 3) after loading the foundation to a prototype building load of 800 kN. After reaching this prototype building load but before increasing the temperature, the foundations were both observed show some creep settlement. A force-displacement feedback loop was used to maintain the same load on the foundation. A greater amount of creep was observed in Test 3 than in Test 2, which may indicate that the soil was slightly softer beneath this foundation. Nonetheless, the foundations all had similar initial slopes to their load-settlement curves. After stabilization under the building load, one of the foundations (Test 2) was heated to 50 °C (the centrifuge temperature was constant at 15 °C) then loaded to failure. The other foundation (Test 3) was heated to 50 °C, cooled down to 20 °C, then loaded to failure.

Rosenberg (2010) performed a detailed soil-structure evaluation of these tests to evaluate the relative contributions of end-bearing and side shear to the ultimate capacity. For simplicity, the ultimate capacities of the foundations can be evaluated in this paper using Davisson's criterion:

$$(1) \qquad Q_{ult} = 0.0038 \text{ m} + 0.01D + QL/AE$$

were D is the foundation diameter in prototype scale and QL/AE is the elastic compression of the foundation. The capacities for Tests 1 through 3 are 1380, 2150, and 1640 kN, respectively. Although the strain distributions with length in the foundation were not measured as in the studies of Bourne-Webb et al. (2009) and Laloui and Nuth (2006), the results in Figure 8 can be explained by the mechanisms noted in their studies (and in Figure 6). The foundation that was heated then loaded to failure (Test 2) had a capacity that was 1.6 times greater than that of the baseline case. This is due to both consolidation of the soil at the tip of the foundation, as well as an increase in horizontal stresses and side friction along the length of the foundation. Although the foundation that was heated, then cooled, then loaded to failure (Test 3) had a capacity that was 1.2 times greater than that of the baseline case it was not cooled below ambient temperatures (which would have caused a relative contraction). The difference between the capacities of the foundations in Tests 2 and 3 can be described by the fact that the foundations both expand during heating, causing consolidation of the soil at the tip of the foundation and along the sides of the foundation. After the foundation in Test 3 is cooled, the horizontal stresses will be less, leading to a lower side shear stress than in Test 2. However, the end bearing should be similar to that in Test 2 (and greater than in Test 1) because of the stiffer soil at the tip of the foundation.

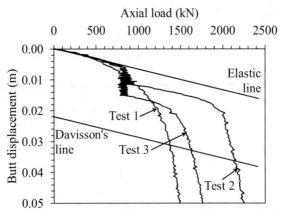

FIG. 9: Load-Settlement curves for three scale-model foundations in prototype scale (Test 1: Baseline loading at 15 °C; Test 2: Heating to 50 °C then loading; Test 3: Heating to 50 C°, Cooling to 20 °C then loading)

In Colorado, most drilled shaft foundations are socketed into rock, which will not change significantly in end bearing. This implies that changes in side friction may still occur, but will likely not influence the performance of the foundation. However, the temperature-induced stresses in thermo-active foundations socketed into rock may be a more important issue to consider due to the rigid end restraint conditions. However, further research is needed to understand this issue.

CONCLUSIONS

A review of the literature on thermo-active foundations indicates that they are a feasible technology to improve the energy efficiency of heating and cooling systems for buildings in Colorado. Further, a comparison of the energy usage of conventional heating and cooling systems with thermo-active foundations indicates that Colorado is an optimal climate for heat exchange technology. This is due to similar heating and cooling demands in the state and the ability for a heat pump system to serve the function of both a furnace and an air conditioner. The importance of evaluating thermo-mechanical effects on the deformation and capacity of foundations was emphasized using results from the literature and from a series of centrifuge scale-model tests. The centrifuge scale-model tests showed a relatively large change in the load-settlement curves for foundations which have undergone different heating and cooling histories. Foundations that were loaded to a target value, heated from 15 °C to 50 °C, then loaded to failure experienced an increase in capacity of 1.6 times above a baseline foundation tested at ambient temperature. However, a foundation heated using the same approach but then subsequently cooled to 20 °C before loading to failure had a capacity that was 1.2 times greater than that of a baseline foundation. These changes in capacity were found to be consistent with mechanisms suggested in the technical literature. The results from these preliminary tests suggest that thermo-active foundations may not lead to significant soil-structure interaction problems in practice. However, further testing is required to develop design guidelines and safety factors for the structural capacity of thermo-active foundations.

ACKNOWLEDGMENTS

The authors appreciate the support of the National Science Foundation grant CMMI-0928159. The views in this paper are those of the authors alone.

REFERENCES

Abu-Hamdeh, N.H. and Reeder, R.C. (2000). "Soil Thermal Conductivity: Effects of Density, Moisture, Salt Concentration, and Organic Matter." Soil Science Society of America. 64:1285-1290.

Adam, D. and Markiewicz, R. (2009). "Energy from earth-coupled structures, foundations, tunnels and sewers." Géotechnique. 59(3), 229–236.

Amis, T., Bourne-Webb, P.J., and Amatya, B. (2008). "Geothermal Business Buoyant." Geodrilling International. Issue 148.

Boënnec, O. (2009). "Piling on the Energy." Geodrilling International.

Bourne-Webb, P. J., Amatya, B., Soga, K., Amis, T., Davidson, C. and Payne, P. (2009). Energy Pile Test at Lambeth College, London: Geotechnical and Thermodynamic Aspects of Pile Response to Heat Cycles. Geotechnique 59(3), 237–248.

Brandl, H. (2006). "Energy Foundations and other Thermo-Active Ground Structures." Géotechnique. 56(2), 81-122.

Choi, J.H. and Chen, R.H.L. (2005). "Design of Continuously Reinforced Concrete

Pavements Using Glass Fiber Reinforced Polymer Rebars." Publication No. FHWA-HRT-05-081. Washington, D.C.

Ebnother, A. (2008). "Energy Piles: The European Experience." GeoDrilling 2008. Ground Source Heat Pump Association (GSHPA). April 30th-May 1st.

Energy Information Agency (EIA). (2008). Annual Energy Review. Report No. DOE/EIA-0384(2008).

Hughes, P.J. (2008). Geothermal (Ground-Source) Heat Pumps: Market Status, Barriers to Adoption, and Actions to Overcome Barriers. Oak Ridge National Laboratory Report ONRL-2008/232.

Kavanaugh, S., Rafferty, K., and Geshwiler, M. (1997). "Ground-Source Heat Pumps – Design of Geothermal Systems for Commercial and Industrial Buildings." ASHRAE. 167 pp.

Krarti, M. and Studer, D. (2009). "Applicability to Carbon and Peak Summer Demand Reduction in Residential Colorado Applications." Report to the Colorado Governor's Energy Office.

Krarti, M. (1995). "Evaluation of the Thermal Bridging Effects on the Thermal Performance of Slab-On-Grade Floor Foundation." ASHRAE Transactions. 103, part 1.

Laloui, L., Nuth, M., and Vulliet, L. (2006). "Experimental and numerical investigations of the behaviour of a heat exchanger pile." IJNAMG. 30, 763–781.

Laloui, N. and Nuth, M. (2006). "Numerical Modeling of Some Features of Heat Exchanger Pile." Foundation Analysis and Design: Innovative Methods (GSP 153). ASCE. Reston, VA. pp. 189-195.

McCartney, J.S., LaHaise, D., LaHaise, T., and Rosenberg (2010). "Feasibility of Incorporating Geothermal Heat Sinks/Sources into Deep Foundations." ASCE Geotechnical Special Publication 198: The Art of Foundation Engineering Practice. Feb. 20-24, 2010.

Omer, A.M. (2008). "Ground-Source Heat Pump Systems and Applications." Renewable and Sustainable Energy Reviews. 12(2), 344-371.

O'Neil, M.W. and Reese, L.C. (1999). Drilled Shafts: Construction Procedures and Design Methods, FHWA Publication No. FHWA-IF-99-025, Federal Highway Administration, Washington, D.C.

Ooka, R., Sekine, K., Mutsumi, Y., Yoshiro, S. SuckHo, H. (2007). "Development of a Ground Source Heat Pump System with Ground Heat Exchanger Utilizing the Cast-in Place Concrete Pile Foundations of a Building." EcoStock 2007. 8 pp.

Rees, S.W., Adjali, M.H., Zhou, Z., Davies, M. and Thomas, H.R. (2000). "Ground Heat Transfer Effects on the Thermal Performance of Earth-Contact Structures." Renewable and Sustainable Energy Reviews. 4(3), 213-265.

Shonder J.A. and Beck, J.V. 1997. "A New Method to Determine the Thermal Properties of Soil Formations from In Situ Field Tests." Oak Ridge National Laboratory Report. ORNL/TM-2000/97. 40 p.

Tarnawski, V. R., Momose, T. & Leong, W. H. (2009). Estimation of quartz content in soils from thermal conductivity data. Géotechnique. 59(4), 331–338.

Wood, C. J., Liu, H. and Riffat, S. B. (2009). Use of Energy Piles in a Residential Building, and Effects on Ground Temperature and Heat Pump Efficiency. Géotechnique 59(3), 287–290.

Helical Pile Capacity to Torque Ratios, Current Practice, and Reliability

James A. Cherry [1] ASCE, P.E. and Moncef Souissi[2], M.S.

[1]Structural Department Manager and IAS Testing Lab (TL #342) Quality Manager, CTL Thompson, Inc, 351 Linden Street, Fort Collins, Colorado 80524; jcherry@ctlthompson.com
[2]IAS Testing Lab Technical Manager, CTL Thompson, Inc, 351 Linden Street, Fort Collins, Colorado 80524; msouissi@ctlthompson.com

ABSTRACT: Historic helical pile ultimate capacities and capacity to torque ratios (Kt) were based upon full-scale tension load tests where ultimate load was taken at deflections near plunge. In 2005, nine helical pile manufacturers joined together in an effort to standardize evaluation requirements for manufacturers seeking approval under International Code Council Evaluation Service (ICC-ES). The result of their efforts was the creation of *"Acceptance Criteria for Helical Foundations Systems and Devices"* (AC358). AC358 now provides standardized guidelines for evaluation of helical piles that includes a net deflection limit at ultimate load equal to 10% of the average helix diameter.

CTL Thompson, Inc. has performed 93 full-scale compression and 109 full-scale tension load tests conducted on behalf of five different helical pile manufacturers seeking ICC-ES evaluation reports under the criteria set forth by AC358. Measured ultimate capacities (Qm) were compared to calculated ultimate capacities (Qu) using AC358 capacity to torque ratios, bearing of the individual blades, and cylindrical shear methods. Statistical analysis of these results shows that helical piles in compression designed with the least Qu determined from torque ratios (Kt), bearing of individual blades, and cylindrical shear methods have a high reliability for all blade configurations. Whereas helical piles in tension based upon the least Qu determined from torque ratios, bearing of individual blades, and cylindrical shear methods have varying reliability depending on blade configuration. Therefore, more study is necessary to determine reliability in tension.

INTRODUCTION

Torque Correlations

Installation torque is widely accepted as one method for predicting helical pile capacity. Hoyt and Clemence, 1989, published a paper relating the ultimate helical pile capacity to torque. Hoyt and Clemence compared ultimate capacities determined from ninety-one multi-helix tension load tests to capacities predicted by torque, individual bearing, and cylindrical shear. In most cases, the ultimate load was taken when an imposed continuous deflection rate of approximately four inches per minute was reached. This deflection rate is near plunging load.

Hoyt and Clemence used the following equation for calculating capacity based upon installation torque:

$$Qu = Kt \times T \qquad\qquad\qquad (1)$$

Where: Qu   =   ultimate capacity
            Kt   =   empirical factor
            T   =   average installation torque

Hoyt and Clemence averaged the installation torque over a distance of penetration equal to three times the largest helix size. Empirical Kt factors used were equal to 33 meters $^{-1}$ (10 feet $^{-1}$) for all shafts less than 89 millimeters (3.5 inches) diameter, and 23 meter $^{-1}$ (7 feet $^{-1}$) for 89 millimeter (3.5 inch) diameter shafts.

Current Practice

The 2009 International Building Code (2009 IBC), specifically section 1810.3.3.1.9, mentions helical piles for the first time. This section states the following:

> *1810.3.3.1.9 Helical piles.*   The allowable axial design load, P_a, of helical piles shall be determined as follows:
>
> $$P_a = 0.5\, P_u \qquad\qquad\qquad (Equation\ 18\text{-}4)$$
>
> Where P_u is the least value of:
>
> 1. Sum of the areas of the helical bearing plates times the ultimate bearing capacity of the soil or rock comprising the bearing stratum.
> 2. Ultimate capacity determined from well-documented correlations with installation torque.
> 3. Ultimate capacity determined from load tests.
> 4. Ultimate axial capacity of pile shaft.
> 5. Ultimate axial capacity of pile shaft couplings.
> 6. Sum of the axial capacity of helical bearing plates affixed to pile.

It can be seen above that allowable helical pile load is defined as one-half of ultimate load (i.e. factor of safety = 2.0). Items 1-3 above relate to the geotechnical capacity of the pile and items 4-6 relate to the structural capacity of the pile. Furthermore, IBC section 1810.1.1 requires deep foundations be designed in accordance to a geotechnical investigation. This geotechnical investigation is necessary to perform the capacity check shown in item number one above. For item number two, the designer must have "well-documented" torque correlations and then torque is monitored during pile installation. These torque correlations historically came from the manufacturers without standardized test methods.   IBC section 1810.3.3.1.2 requires full-scale load testing if a reduced factor of safety is used for design purposes, when pile capacities are in doubt, or when required by the building official. The least value of these capacity checks is the pile rated capacity.

In 2005, nine helical pile manufacturers worked together in an effort to standardize helical pile evaluation requirements for manufacturers seeking approval under International Code Council Evaluation Service (ICC-ES). The result of their efforts was the creation of *"Acceptance Criteria for Helical Foundations Systems and Devices"* (AC358).   AC358 requires manufacturers to perform various standard calculations and testing.   AC358 testing includes torsion capacity, helix flexure capacity, coupling rigidity, and full-scale field load tests.   The full-scale field load tests are necessary to verify the Kt values for each manufacturer and individual shaft sizes.   AC358 provides the conforming Kt values shown in Table 1.

Table 1. AC358 Conforming Kt Values

Shaft Size millimeters (inches)	Kt meter $^{-1}$ (feet $^{-1}$)
38 (1.5)	33 (10)
44 (1.75)	33 (10)
73 (2.875)	30 (9)
76 (3.0)	26 (8)
89 (3.5)	23 (7)

For full-scale field load tests, AC358 requires manufacturers to perform a minimum of eight compression and eight tension load tests per shaft size in order to verify the Kt values presented in Table 1.   Testing includes at least one compression test at maximum torque, and one tension test at maximum torque on the smallest and largest single helix size per shaft size. Furthermore, AC358 defines ultimate load when the net (total minus elastic) deflection equals 10% of the average helix diameter. This limit in maximum deflection generally reduces the ultimate capacity for the same pile had ultimate load been measured near plunging load.

Figure 1 shows results from a typical compression load test. It can be seen that if ultimate load is measured near plunge, the pile ultimate capacity would have been around 80 kips. Whereas at a deflection limit of 0.8 inches (10% of an 8 inch helix), the ultimate capacity would be approximately 72 kips.

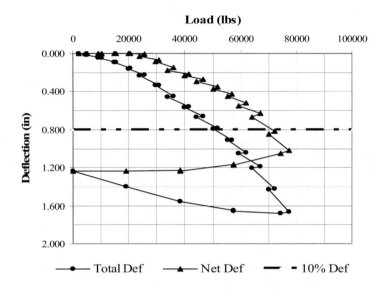

FIG. 1. Typical Compression Load Test Curve

AC358 requires measured ultimate capacities (Qm) be compared to calculated allowable capacities (Qa) with Factor of Safety = 2.0, and using installation torque and equation (1). AC358 defines a helical product to be conforming, and the Kt values shown in Table 1 deemed valid; if the average ratio of Qm/Qa exceeds 2.0 and if none of the load tests have a Qm/Qa ratio less than 1.0. If the average ratio of Qm/Qa falls below 2.0 or one Qm/Qa falls below 1.0, then the manufacturer is required to perform additional load tests to obtain a unique Kt value for their product.

DATA COLLECTION

Test Sites

The 202 load tests used for this study were performed at seven different test sites. Five of the test sites were located in the northern Colorado area near CTL Thompson's Fort Collins office, one test site was located in Westchester, Ohio, and one test site was located in Lander, Wyoming. Of the seven test sites, a vast majority of the testing took place at the CSU-1, Windsor-1, and Windsor-2 sites. Soil test data for each site consisted of typical soil investigations. This included soil borings with blow counts at 5-10 foot intervals and gradations.

The CSU-1 test site generally consists of a 12 to 18 foot deep very stiff to hard sandy to slightly sandy clay over medium hard weathered and unweathered claystone bedrock. Cohesion of the clay ranged from 3000 psf to 3500 psf and for the claystone bedrock ranged from 3500 psf to 6000 psf.

The Windsor-1 test site generally consists of a 15 to 19 foot deep loose to dense silty sand with gravel over moderately hard to hard weathered and unweathered sandstone/siltstone bedrock. The friction angle ranged from 30 to 40 degrees. The Windsor-2 test site generally consists of a 15 to 17 foot deep very loose to loose sand with occasional gravel over 11-12 feet of medium to dense gravely sand with occasional cobbles. The friction angle ranged from 30 to 35 degrees

Load Test Data

Load test data from five different helical pile manufacturers consisting of 93 full-scale compression and 109 full-scale tension load tests were analyzed. The load testing was performed by representatives of CTL Thompson, Inc. as part of an AC358 testing program for each manufacturer. Pile sizes ranged from 38 millimeter (1.5 inch) square bars to 89 millimeter (3.5 inch) diameter shafts. Helix sizes ranged from 203 millimeters (8 inches) to 355 millimeters (14 inches). Helix pitch was typically near 76 millimeters (3 inches). Piles consisting of single and multi-helix configurations were tested at seven different test sites with soil strata consisting of clay, sand, and weathered bedrock. Installation depths ranged from 1.83-12.5 meters (6-41 feet). Tension tests had a minimum length or embedment from grade equal to at least 12 times the upper most helix diameter. A summary of the pile types and number of tests performed is provided in Table 2.

Table 2. Summary of Pile Types and Load Tests

Shaft Size millimeters (inches)	Helix Configuration	Compression Tests	Tension Tests	Total Tests
38 (1.5)	Single	9	8	17
	Multi	10	8	18
44 (1.75)	Single	4	5	9
	Multi	4	4	8
73 (2.875)	Single	14	20	34
	Multi	13	19	32
76 (3.0)	Single	12	15	27
	Multi	9	5	14
89 (3.5)	Single	14	18	32
	Multi	4	7	11

All load testing was performed in general accordance with AC358 and ASTM D-1143-81(1994)e1 Standard Test Methods for Deep Foundations Under Static Axial Compressive Load and ASTM D-3689-90(1995) Standard Test Methods for Deep Foundations Under Static Axial Tensile Load. All test piles were installed relatively vertical, and in all cases, the quick load test method was used. Installation torque was measured by differential pressure gages attached to a calibrated (traceable to NIST) torque motor. Torque was recorded as the piles advanced and at termination depth.

Final torque was measured over the last few revolutions of the helical pile while making sure the pile advanced at least 85% of the helix pitch. An example of typical compression and tension load frame setups are shown in Figures 2 and 3 respectively.

FIG. 2. Typical Compression Test Load Frame Setup

FIG. 3. Typical Tension Test Load Frame Setup

Calculated Capacities

Capacities based upon measured installation torque, area of the helices times ultimate bearing capacity, and cylindrical shear were calculated. For the purposes of this study, the procedures used for calculating ultimate bearing capacity and cylindrical shear were as presented in the text "Helical Piles A Practical Guide to Design and Installation" (Perko 2009). Methods described by Perko were used to estimate friction angles, and cohesion.

RESULTS

Load test data was grouped by shaft size, single or multi-helix, and by compression or tension. The least calculated ultimate capacity by torque, bearing, or cylindrical shear (Qu) was compared to the measured ultimate load test capacity (Qm) for each test. Statistical analysis of the results was performed using Easyfit™ a statistical software program. Cumulative distribution functions were generated using the Qm/Qu ratios. The software program, using the Anderson-Darling test method, generated best-fit probability functions. The Anderson-Darling test method was used because it puts more emphasis on the tails of the functions than other test methods available in the software. Another criteria for selecting the best fit curve was that the Qm/Qu ratio could not be less than zero. An example output of a probability density function is shown in Figure 4. Figure 4 shows the histogram results in the background along with the probability density function. The Y axis shows the probability at a given value and the X axis shows the ratios of Qm/Qu.

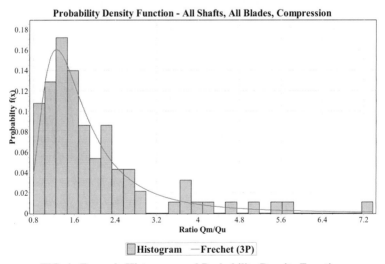

FIG. 4. Example Histogram and Probability Density Function

The least Qu determined from torque, bearing of individual blades, and cylindrical shear methods were compared to Qm. These results were deemed acceptable if the probability of the ratio of Qm/Qu exceeded 0.5 97.7% of the time (Perko 2009, Duncan 2000). The authors acknowledge that higher reliability may be required on certain projects such as federal highway projects or other essential facilities. A lower bound ratio of 0.5 for Qm/Qu was selected to measure reliability because a minimum factor of safety of 2.0 is required in helical pile design per the 2009 IBC and AC358.

Results of the study are presented in Tables 3 and 4. In some cases, insufficient data was available to obtain probability density functions that visually fit the histograms well. In these cases, the reliability calculation has been omitted.

Table 3. Summary of Qm/Qu Ratios

Shaft Size millimeters (inches)	Comp. or Tension	Blade Config.	Qm/Qu Min	Qm/Qu Max	Std. Dev	Qm/Qu Mean
38 (1.5) and 45 (1.75) square	C	Single	0.99	4.59	1.14	2.01
		Multi	1.00	3.15	0.54	1.58
		All	0.99	4.59	0.89	1.79
	T	Single	0.62	3.23	0.82	1.55
		Multi	0.87	1.73	0.27	1.30
		All	0.62	3.23	0.62	1.43
73 (2.875) φ	C	Single	0.80	5.08	1.32	2.14
		Multi	0.86	2.85	0.53	1.41
		All	0.80	5.08	1.07	1.79
	T	Single	0.58	3.31	0.84	1.59
		Multi	0.53	1.87	0.35	1.01
		All	0.53	3.31	0.71	1.31
76 (3.0) φ	C	Single	1.14	5.41	1.16	2.10
		Multi	0.87	1.68	0.27	1.34
		All	0.87	5.41	0.96	1.78
	T	Single	0.69	2.84	0.76	1.63
		Multi	1.18	1.55	0.16	1.34
		All	0.69	2.84	0.67	1.56
89 (3.5) φ	C	Single	0.83	2.60	0.63	1.74
		Multi	1.05	4.46	1.77	2.85
		All	0.83	4.46	1.04	1.99
	T	Single	0.91	4.33	1.17	2.31
		Multi	0.86	2.39	0.63	1.47
		All	0.86	4.33	1.1	2.08

Table 4. Summary of Statistical Analysis

Shaft Size millimeters (inches)	Compression or Tension	Blade Configuration	Best Fit Curve	Calculated Reliability[1]
ALL	C	Single	Log Pearson (3P)	0.999
		Multiple	Burr 4P	0.999
		All	Frechet (3P)	0.999
	T	Single	Johnson SB	0.986
		Multiple	Burr	0.986
		All	Fatigue Life	0.999
73 (2.875) φ	C	Single	Johnson SB	0.999
		Multiple	Frechet (3P)	0.999
		All	Fatigue Life (3P)	0.999
	T	Single	Johnson SB	0.964
		Multiple	n.a.[2]	n.a.[2]
		All	Fatigue Life (3P)	0.999

1. reliability that Qm/Qu exceeds 0.5 (i.e. FS=2.0)
2. n.a. indicates more data needed

CONCLUSIONS

In general piles performed better in compression than in tension. For all compression tests the average Qm/Qu equaled 1.82 whereas for tension the average Qm/Qu equaled 1.56. In compression, taking the least calculated Qu as described earlier in current practice, resulted in a 99.99% reliability when using a FS = 2.0. In other words, there is a 99.99% chance that the pile deflection will not exceed 10% of the average helix diameter at calculated allowable load. Reliability may be different if the deflection criteria is changed.

In tension, results were found to be variable. Some results suggest a high reliability whereas some suggest a lower reliability. This variability could be explained by the following reasons. First, historically measured ultimate capacity (Qm) was taken at or near plunging load, whereas, AC358 limits deflection at ultimate capacity to 10% of the average helix diameter. Secondly, piles penetrating harder layers may not be embedded far enough to prevent pull-out within that layer. Thus, more data and further study is necessary to determine reliability in tension. As more data is collected using criteria set forth in AC358, statistical analysis trends should become clearer.

ACKNOWLEDGMENTS

The Authors would like to thank CTL Thompson, Inc. for its support in preparation of this paper. In addition, the Authors would like to thank Becky Young for editing and the manufacturers who participated in the AC358 testing program for allowing use of the data.

REFERENCES

Hoyt, R.M., Clemence, S.P., (1989). "Uplift Capacity of Helical Anchors in Soil". *Proceedings of the 12th International Conference on Soil Mechanics and Foundation Engineering, Vol. 2, pp. 1019-1022.*

Perko, H.A., PhD, PE, (2009). "Helical Piles - A Practical Guide to Design and Installation"

Duncan, J.M., (2000). "Factors of Safety and Reliability in Geotechnical Engineering". *Journal of Geotechnical and Geoenvironmental Engineering.*

AC358, (2007). "Acceptance Criteria for Helical Foundation Systems and Devices" International Code Council Evaluation Service (ICC-ES).

International Building Code (IBC), (2009), International Code Council

Case Study: Heave Potential Associated with Ettringite Formation in Lime Treated Materials for an Aurora, Colorado Roadway

Darin R. Duran[1], P.E.

[1]Principal, Geotechnical Engineering Manager, Joseph A. Cesare and Associates, Inc. 7108 South Alton Way, Unit B, Centennial, Colorado, 80112; dduran@jacesare.com

ABSTRACT: A common technique to improve the performance of a roadway with clay subgrade is to mix the clay with lime. The lime reacts with the clay minerals to reduce the plasticity of the clay and increase the strength to allow for a reduction in the pavement section thickness. When sulfates are present in the subgrade in sufficient quantities, the lime can react with the sulfates to create secondary minerals that can cause heaving.

In a high sulfate environment, the risk associated with heave due to the formation of secondary minerals can be reduced by applying a double application of lime. After the first application, the soil/lime mixture is cured to allow the minerals to form. The second application is then applied and the mixture is compacted to gain the benefits of lime treated clay subgrade with a reduced risk of heave. Part of a geotechnical study conducted for roadways surrounding an educational campus in Aurora, Colorado included a heave analysis of lime treated clay subgrade in a high sulfate environment. The results indicated a required initial mellowing period of 16 days with 3.3 percent volume increase. After the second application the lime treated clay subgrade continued to heave for another 23 days with 3.5 percent volume increase.

INTRODUCTION

Swelling clay minerals are common along the Colorado Front Range. Commercial, public and residential development frequently encounters these potentially swelling clays. Therefore a variety of mitigation techniques have been developed by geotechnical engineers to reduce the risk associated with structures bearing on swelling materials. For roadway construction a common technique is to excavate materials to a certain depth and re-compact the materials at a higher moisture content. This homogenizes the subgrade materials, reducing the risk of differential swell. It reduces the swell potential by increasing the moisture content and decreasing the permeability by removing layering, fractures or other structure in the subsurface that can provide a conduit for water flow.

The moisture contents during compaction are typically zero to four percent above optimum moisture content as determined from a standard Proctor. This creates a potentially unstable subgrade condition for the placement of pavement. Stabilizing the subgrade with lime can improve its performance. The lime reacts with the clay to reduce its plasticity, swell potential, and increase its strength. The increase in strength can allow for a reduction in the pavement section thickness.

Lime is a high pH, calcium based product and in the presence of soluble sulfates can form secondary minerals, one of which is Ettringite (Little, 1995). Ettringite is a hydrous calcium alumino-sulfate mineral and forms in a high pH environment (Little and Nair, 2009). The formation of Ettringite is problematic because of the expansion that occurs during the formation of the crystal. In low concentrations of soluble sulfates (less than 0.3 percent) the risk associated with Ettringite formation are low. Risk associated with soluble sulfate contents greater than 1.0 percent are considered unacceptable (Little and Nair, 2009). Locally, concentrations of 0.2 percent and greater (CDOT, 2007 and MGPEC, 2007) are generally considered the threshold where the geotechnical engineer should consider the potential effects of soluble sulfates when calcium is present in the stabilizing agent.

A common technique in a soluble sulfate environment is to mix the subgrade with lime and water and allow the Ettringite to form prior to compacting the mixture. Frequently the lime is mixed in a double application where half the lime is added and the mixture is allowed to cure until expansion from Ettringite is complete then the second half of the lime is applied prior to compaction. This paper describes a project where a mix design was performed utilizing a double application of lime in a high soluble sulfate material.

PROJECT BACKGROUND

The construction of an educational campus in eastern Aurora, Colorado included the construction of new roadways within City of Aurora right-of-ways to provide access to the campus facilities. The campus is located at the southwest corner of East 6th Avenue and South Harvest Road. At the time of construction East 6th Avenue and South Harvest Road consisted of two lane arterials paved with hot mix asphalt with the exception of the intersection of East 6th Avenue and South Harvest Road which was paved with Portland cement concrete.

The new construction consisted of the widening of East 6th Avenue to provide a turn lane onto southbound South Harvest Road, two southbound lanes of South Harvest Road and the extension of East 1st Avenue, a two lane collector road, west of South Harvest Road. A site plan is shown on Figure 1.

Grading on South Harvest Road and East 1st Avenue required cuts of up to 2.4 meters (8 feet) with fills of up to 1 meter (3 feet). Little grading was required for the East 6th Avenue widening. All grading utilized on-site soil.

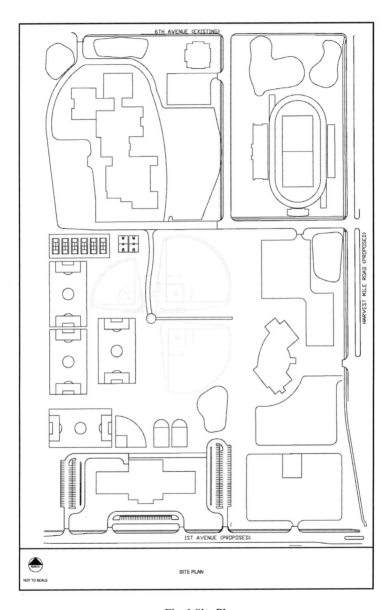

Fig. 1 Site Plan

Subsurface Study

Studies included an evaluation of the subgrade to support the new pavement sections. For the evaluation, the subgrade soils were drilled and sampled along centerline of the proposed roadway construction. Test holes were spaced at one per 61 linear meters (200 linear feet) along East 6th Avenue and South Harvest Road and one per 76 linear meters (250 linear feet) on East 1st Avenue. Depth of the drilling varied between 1.5 meters (5 feet) and 3 meters (10 feet). Representative bulk samples of the subgrade material were taken from subgrade level to a depth of 1.2 meters (4 feet) below subgrade level. At 1.2 meters, samples of the soil were taken using a modified California sampler, which is driven into the soil by dropping a 63.5 kg (140-pound) hammer through a free fall of 76 centimeters (30 inches). The modified California sampler is a 6.4 centimeter (2.5 inch) outside diameter by 5.1 centimeter (2 inch) inside diameter barrel sampler. The barrel is 30.5 centimeters (12 inches) in length with four 7.6 centimeter (3 inch) long brass tube liners inserted into the barrel. Soil samples are collected within the brass liners during sampling.

Laboratory Testing

Samples were returned to the laboratory where they were visually classified and tested for various engineering properties. Testing included grain size distributions, Atterberg limits, swell/consolidation potential and soluble sulfate content (HACH Method). Swell/consolidation tests were performed on select samples collected in the brass liners to evaluate their swell potential upon wetting at a surcharge load of 0.3 kilopascals (200 pounds per square foot). A summary of the results of testing on each roadway is shown in Table 1.

TABLE 1. Summary of Laboratory Test Results

Roadway	Soil Description	AASHTO Soil Classification	USCS Soil Classification	Swell Potential (%)	Soluble Sulfate (%)
East 6th Avenue	Sandy lean clay	A-6	CL	0.3 to 1.4	0.05 to 0.69
South Harvest Road	Sandy lean clay	A-4 to A-6	CL	-0.6 to 12.9	<0.01 to 1.77
East 1st Avenue	Clayey sand to Sandy lean clay	A-4 to A-6	SC to CL	3.5 to 15.7	<0.01 to 1.53

Swelling Subgrade Mitigation

Since the roadways are within the right-of-ways of the City of Aurora, The City of Aurora Roadway Design and Construction Specifications, 2006 edition (Specifications), governed the design of the pavement sections and construction of the subgrade. Table 5 in Section 5.0 of the Specifications requires a minimum of 0.5 meters (1.5 feet) of moisture treatment and a minimum of 0.3 meters (1 foot) of chemical stabilization if the swell potential of subgrade exceeds five percent. East 6th Avenue falls below this threshold but South Harvest Road and East 1st Avenue did not.

Section 22.0 of the Specifications requires that the chemical stabilization consist of lime or a combination of lime and fly ash when the stabilization is being used to mitigate swell potential. Section 22.07.1.03 High Sulfate Treatment states *"Where sulfates are over 0.5 percent and less than 1.0 percent the Geotech must address the method of treatment"* and *"When a double treatment of lime is required, the first 50 percent of the agent shall be placed, moisture treated and allowed to mellow or cure for up to three weeks, as determined by the Geotech. The last half of the lime shall then be applied."* The specification does not give recommendations, requirements, or variance to mitigation techniques when sulfate contents exceed 1.0 percent. City of Aurora indicated that they have had success on other projects through double application of lime with sulfates in excess of 1.0 percent. Based on this information, the laboratory study continued to include a lime mix design, employing a double application of lime.

LIME MIX DESIGN

The materials used in the mix design were those that exhibited some of the highest swell potential and soluble sulfate contents encountered during the subsurface study. The properties of the material are presented in Table 2.

TABLE 2. Properties of Subgrade Material Used in Mix Design

-#4 Material	-#200 Material	LL	PI	Water Soluble Sulfate
100%	65.1%	37	24	1.46%

A moisture/density relationship was performed on the material according to standard Proctor (ASTM D 698) to evaluate the maximum dry density and optimum moisture content of the untreated material. The results indicated a maximum dry density of 1,751 kilogram per cubic meter (109.3 pounds per square foot) and 16.4 percent optimum moisture content.

An Eads-Grimm pH test was conducted on the sample in order to evaluate the optimum lime content. The Eads-Grimm pH test involves mixing the soil with various percentages of lime to determine how much lime is required to saturate to soil with lime. The results of the test are shown in Table 3. Based on the results of the test, the sample became saturated with lime at about 4.5 percent of quick lime by dry weight. A base lime content of 5.0 percent quick lime was selected to conduct the mix design. This equates to 6.6 percent by dry weight of hydrated lime.

TABLE 3. Eads-Grimm pH Test

% Quick Lime By Dry Weight	pH
2.0	12.0
3.0	12.2
4.0	12.3
5.0	12.3
6.0	12.4
7.0	12.4
8.0	12.4

The sample was passed through a No. 4 screen. Portions of the sample were mixed with 5.0, 6.0, and 7.0 percent quick lime, moistened to various moisture contents and allowed to cure for a minimum of 24 hours. After 24 hours, the soil/lime mixtures were tested for Atterberg limits and pH. The Atterberg limits of the treated material at each lime content were non-plastic and the pH results were 12.3 indicating the appropriate specified limits of these properties had been met. Moisture/density relationship tests were performed on each of the lime content samples according to ASTM D698. The soil/lime cylinders created from each proctor were wrapped in plastic and moist cured for five days at a temperature of 38 degrees C (100 degrees F). After five days the soil/lime cylinders were then tested to determine their unconfined compressive strength. The testing indicated 5.0 percent quick lime or 6.6 percent hydrated lime was the design lime content. The results of this testing is presented in Table 4.

TABLE 4. Laboratory Test Result Soil/Lime Mixtures

% Lime	Maximum Dry Density kg/m^3 (pcf)*	Optimum Moisture Content (%)*	Unconfined Compressive Strength kPa (psi)**
5	1,654 (103.3)	18.0	53.6 (250)
6	1,623 (101.3)	19.8	58.9 (275)
7	1,621 (101.2)	19.3	78.2 (365)

*Based on ASTM D698, Standard Proctor
**Average of four test specimens rounded to the nearest 5 psi.

Due to the high soluble sulfate concentration in the soils, the remainder of the original soil sample was prepared to evaluate the swell potential from Ettringite formation. The lime was added to the samples in a double application, one-half at a time. At the first lime application, a remolded swell sample was prepared and tested for swell under controlled moisture conditions and 0.3 kilopascals (200 psf) surcharge. The swell was monitored and the remainder of the lime was applied when the swell appeared complete. These results indicated that an initial curing period of 16 days was required for the initial lime application with a swell potential of 3.3

percent. The remainder of the lime was applied to the samples at 16 days. A second remolded swell sample was prepared and tested under the same conditions as the first. After the second application the sample swelled an additional 3.5 percent over a period of 23 days. At the end of the 23 day cure, the sample was tested for soluble sulfates. The soluble sulfate content reduced to an acceptable level of 0.11 percent. Figure 2 shows a plot of the swell versus time after each application of lime.

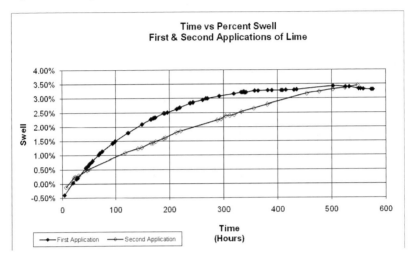

Fig 2. Swell vs. Time after Application of Lime

ATLERNATIVE SUBGRADE CONSTRUCTION

South Harvest Road and East 1st Avenue had a strict deadline for construction to be completed in order to provide fire access to the site during construction of primary and secondary schools on the campus. Due to the deadline, the project schedule could not accommodate a 40 to 45 day construction period for the stabilization of the subgrade. As a result, the City of Aurora allowed an alternative design in lieu of the stabilized subgrade. The alternative design consisted of 0.3m (12 inches) of aggregate base course over a biaxial geogrid. The purpose of the alternative was to provide a stable base for paving with similar strength characteristics.

CONCLUSION

Risks associated with lime stabilization in an environment where the soluble sulfate is greater than 1.0 percent is generally considered unacceptable. On this project soluble sulfates in excess of 1.0 percent were encountered and jurisdictional specifications that required stabilization with lime due to the high swell potential of the subgrade materials. Testing indicated that the risk associated with the soluble sulfates can be remediated through a double application of lime. The remediation, however, may require a curing period that could adversely impact the construction schedule. In a high sulfate environment a mix design should also include testing to determine the length of curing. For this case study the lime stabilization requirement was waived and an alternative subgrade stabilization method was utilized consisting of aggregate base over biaxial geogrid.

REFERENCES

Little, D. N. (1995). "Stabilization of Pavement Subgrades & Base Courses with Lime." Kendall/Hunt Publishing Company, Dubuque, Iowa.
Little, D. N., Nair, S. (2009). "Recommended Practice for Stabilization of Sulfate Rich Subgrade Soils." NCHRP Web-Only Document 145.
Colorado Department of Transportation (2007)."2007 Pavement design Manual."
City of Aurora (2006). "Roadway Design and Construction Specifications."
Metropolitan Government Pavement Engineering Council (MGPEC) (2007). "Pavement Design and Construction Standards."

Utilizing "Grout Columns" for Soil Nail Wall Construction

John H. Hart[1], MSCE, P.E., James V. Warnick, Jr.[2], MSCE, P.E.

[1]Engineer, Coggins and Sons, 9512 Titan Park Circle, Littleton, CO 80125; phone (303) 791-9911; FAX (303) 791-0967; jhart@cogginsandsons.com

[2]Engineer, Coggins and Sons, 9512 Titan Park Circle, Littleton, CO 80125; phone (303) 791-9911; FAX (303) 791-0967; jwarnick@cogginsandsons.com

ABSTRACT: It is well known that earth retention incorporating soil nail construction is unfavorable in soil conditions consisting of sand, gravels, cobbles, and boulders; soils typically encountered in Colorado mountain regions. With the use of "grout columns", a soil nail earth retention system can be utilized in these soil conditions providing a safe method of earth retention while maintaining an efficient construction schedule.

Construction, rationale, and benefits of incorporating grout columns in soil nail wall construction are discussed. Specific focus is on "arching" of the soil between grout columns to reduce the loss of soil, enhanced vertical support of the shotcrete face provided by the grout columns while excavating to the next level, and wall movements from a soil nail earth retention project from the Colorado mountain region incorporating grout columns.

INTRODUCTION

Soil nail earth retention is typically used to stabilize excavations where top-down construction is advantageous relative to other retaining wall systems. Soil nail earth retention construction consists of passive reinforcement (no post tensioning) of existing ground by installing closely spaced high strength steel bars (nails), which are subsequently encased in grout. A high strength steel bar encased in grout constitutes a soil nail. The soil nails are installed into the exposed excavation face with an inclination of typically 10 to 20 degrees below horizontal and are primarily subjected to tensile stresses. As excavation proceeds, shotcrete is applied to the exposed excavation face to provide horizontal support (FHWA 2003).

Soil nail earth retention requires excavation cuts of typically 1.5 m (5 ft) in height. The soil must have some degree of "cohesion" in order to utilize soil nail earth retention. The exposed excavation face must stand unsupported, prior to soil nail installation and shotcrete application. Therefore, soil nailing, is not well suited in all geotechnical conditions (FHWA 1996).

Examples of unfavorable or difficult geotechnical conditions for soil nail earth retention are dry, poorly graded cohesionless soils, soils with excessive moisture / wet pockets, and soils containing cobbles, and boulders.

Dry cohesionless soils tend to ravel when exposed and "run" from behind the shotcrete when proceeding to the next excavation level. This causes the shotcrete to become unsupported and the supporting nail at the face experiences tensile stresses and greater bending stresses.

Soils containing wet pockets tend to slough and create face stability problems. In addition, groundwater seeping to the exposed excavation face may cause difficulties for shotcrete adhesion.

Cobbles and boulders present a challenge to the excavation process and/or excavator. Cobbles and boulders encountered at the excavation face must be removed by the excavator. Removing the obstacle (cobble or boulder) can jeopardize the previously placed shotcrete or cause the soil behind the obstacle to loosen and ravel. This causes the shotcrete to become unsupported, or can result in large shotcrete quantity overruns by filling the void created.

All before-mentioned unfavorable geotechnical conditions exist in Colorado mountain regions. A soil nail earth retention system utilizing grout columns can be constructed in these geotechnical conditions to provide a safe and redundant method of earth retention. Rotary percussion drilling systems utilized for grout column and soil nail installation also maintain an efficient and predictive construction schedule.

CONSTRUCTION SEQUENCE

The typical construction sequence of soil nail earth retention incorporating grout columns consists of five steps. The construction sequence is described further and shown schematically in Figure 1.

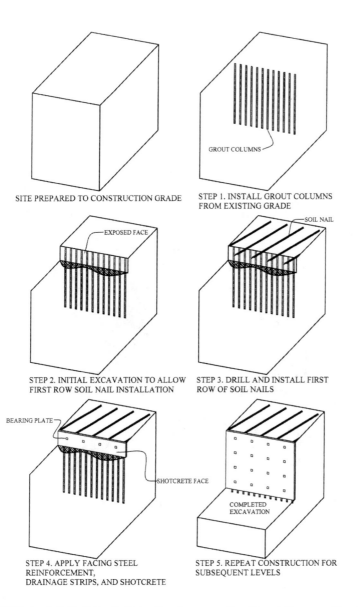

FIG. 1. Soil Nail Earth Retention Incorporating Grout Columns Construction Sequence

Step 1. Install Grout Columns

Grout columns are installed at the beginning of earth retention construction along the perimeter of proposed excavation immediately behind the face of earth retention. The grout columns are typically 152 mm (6 in) in diameter and are spaced 0.61 m (2 ft) on center.

The excavation perimeter is determined at the construction site and marked on the ground. Accurate location of the grout columns relative to the proposed construction is critical to ensure construction of the earth retention system does not impede the proposed construction.

After the grout column layout is accomplished, the grout columns are drilled. Typically, rotary percussion drilling systems as shown in Figure 2 are utilized in the geotechnical conditions described earlier. The grout columns are drilled from existing grade to a predetermined depth below the bottom of the proposed excavation, typically 0.31 m (1 ft) below the proposed bottom of excavation.

FIG. 2. Drilling Grout Columns for Soil Nail Earth Retention

Once the desired depth is obtained, a high strength steel bar with a tremie tube attached is inserted. On-site-prepared high strength grout is pumped into the drill hole from the bottom to the top. The grout typically has a water/cement ratio ranging from 0.4 to 0.5 and exhibits a minimum 28-day unconfined compressive strength of 28 MPa (4,000 lb/in^2). It is desirable to install all grout columns around the excavation perimeter before excavation occurs.

Step 2. Initial Excavation

The depth of the initial excavation lift is typically 1.5 m (5 ft) and extends below the elevation where the first row of nails will be installed. The initial excavation exposes the previously placed grout columns as shown in Figure 3.

FIG. 3. Initial Excavation and Exposed Grout Columns

At times, the initial excavation extends deeper, 2.2 m to 2.4 m (7 ft to 8 ft); however, shotcrete will be placed to cover the upper 1.5 m (5 ft). For example, the deeper initial excavation is utilized where soil nail installation would interfere with a shallow utility. The grout columns allow a deeper cantilever section to the top row of soil nails, permitting the soil nails to pass under the shallow utility without interference.

Step 3. Drill and Install First Row of Soil Nails

After excavation exposes the initial lift, soil nail drilling and installation commences. As with the grout columns, rotary percussion drilling systems are typically utilized in the geotechnical conditions encountered in the Colorado mountain regions to install soil nails. The drill holes for soil nails are drilled to a specified length, diameter, inclination, and horizontal spacing.

Once the specified depth of the drill hole is obtained, a high strength steel bar, 520 MPa (75 kips/in^2), with tremie tube and centralizers attached is inserted into the drill hole. High strength grout, prepared on-site, is pumped into the drill hole from the bottom to the top. The grout is commonly placed under low pressure 1 MPa (150 lb/in^2). Figure 4 shows installation of the top row of soil nails.

FIG. 4. Installation of the Top Row of Soil Nails

Step 4. Apply Facing Steel Reinforcement, Drainage Strips, and Shotcrete

The application of facing steel reinforcement and shotcrete transfers the earth loads applied from the existing exposed face to the previously installed soil nails. In addition, geocomposite drainage strips are attached to the exposed excavation face, between nails, to collect and direct any current or future groundwater. The steel reinforcement typically consists of welded wire mesh located in the middle of the shotcrete thickness. Vertical and horizontal reinforcement bars are located at each nail.

After attaching steel reinforcement to the exposed face, shotcrete is applied to a pre-selected thickness. The shotcrete typically has a 28-day compressive strength of 28 MPa (4,000 lb/in^2) and a 3-day compressive strength of 14 MPa (2,000 lb/in^2). While the shotcrete is still "wet", a steel bearing plate is placed over the nail head and lightly pressed into the shotcrete. A hex nut is then attached to the end of the steel bar to secure the bearing plate to the nail (See Figure 5 for Soil Nail Cross Section).

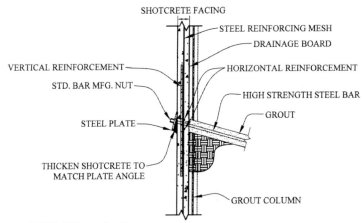

FIG. 5. Soil Nail Cross Section

Step 5. Repeat Construction for Subsequent Levels

Steps 2 through 4 are repeated for the remaining excavation lifts until the bottom of excavation is attained (See Figure 6). The drainage strips extend below each lift and ultimately extend to the bottom of the excavation where they are incorporated into a collection system.

FIG. 6. Drilling Bottom Row of Soil Nails

DESIGN OF GROUT COLUMNS

Grout columns, in general, provide two mechanisms of resistance during excavation. The mechanisms of resistance are in the horizontal and vertical directions. First, the grout columns provide horizontal resistance to the applied earth pressure while excavating from lift to lift. Second, the grout columns provide vertical resistance or additional axial capacity to the soil nail face while excavating from lift to lift. Both mechanisms of resistance, horizontal and vertical, as they relate to the design of grout columns are discussed further.

Horizontal Resistance

The geotechnical conditions encountered in the Colorado mountain regions are typically non-cohesive. These types of soil have a tendency to "flow" while excavation proceeds from lift to lift. Properly spaced grout columns provide horizontal resistance and reduce the potential of "flowing" soils. Basically, grout columns provide the cohesionless soil with an apparent cohesion, allowing the soil to "arch" between the grout columns.

The potential for "flowing" soils must be analyzed to determine the optimum horizontal spacing of the grout columns while providing adequate resistance to the applied earth pressure. This potential can be determined with a procedure shown in (Pearl, Campbell, and Withiam, 1992). See Figure 7.

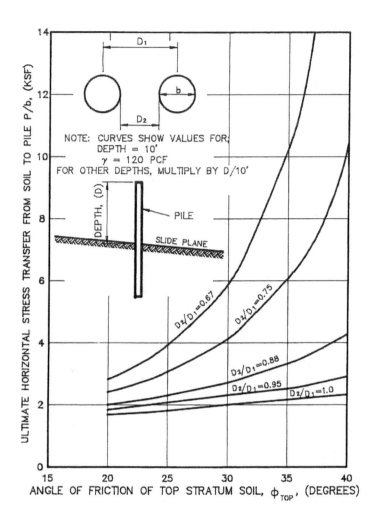

FIG. 7. Ultimate Stress Transfer from Soil to Piles vs. Shear Strength of Soil

Utilizing Figure 7 with $\varphi = 30$ degrees and $\gamma = 19.1$ kN/m^3 (120 lb/ft^3), the Factor of Safety between earth pressure and grout column resistance for various horizontal grout column spacings is shown in Table 1.

Table 1. Factor of Safety for Various Grout Column Horizontal Spacing

b mm (ft)	D_1 mm (ft)	D_2 mm (ft)	D_2/D_1	P (Ult.) kN/m (kips/ft)	P (Ser.) kN/m (kips/ft)	E_p kN/m (kips/ft)	Factor of Safety
152 (0.50)	457 (1.50)	305 (1.00)	0.67	43.8 (3.00)	21.9 (1.50)	6.13 (0.42)	3.57
152 (0.50)	610 (2.00)	457 (1.50)	0.75	29.2 (2.00)	14.6 (1.00)	8.18 (0.56)	1.79
152 (0.50)	762 (2.50)	610 (2.00)	0.80	26.3 (1.80)	13.2 (0.90)	10.2 (0.70)	1.29
152 (0.50)	914 (3.00)	762 (2.50)	0.83	23.3 (1.60)	11.7 (0.80)	12.3 (0.84)	0.95

Notes:

b = Diameter of grout column

D_1 = Center to center distance of grout column

D_2 = Clear space between grout columns

P (Ult.) = Determined from Figure 7

P (Ser.) = P (Ult.) / 2

E_p = Earth Pressure = $(K_a \gamma H) * D_1$; where H = 2.13 m (7 ft) (Maximum excavation lift height throughout the soil nail wall construction.)

Factor of Safety = P (Ser.) / E_p

The geotechnical conditions encountered in the Colorado mountain regions have similar properties to those used for the calculations in Table 1. It can be seen from Table 1 that as the grout column horizontal spacing increases, the Factor of Safety decreases. Therefore, a 0.61 m (2 ft) center to center spacing of grout columns yields an appropriate Factor of Safety (author's opinion; greater than 1.35 for temporary condition).

It is not surprising that closely spaced grout columns provide higher resistance to the applied earth pressure than grout columns spaced farther apart. The horizontal spacing of grout columns should consider cost and constructability while providing an appropriate Factor of Safety.

In addition to determining the "flowing" potential of soil between the grout columns, the structural capacity for horizontal resistance of the grout columns should be determined. The resisting bending moment from the composite section of circular grout in the drill hole and the high strength steel bar can be calculated. This calculated value is then compared to the bending moment calculated from the applied earth pressure. An appropriate Factor of Safety should exist between these two calculated values (author's opinion; greater than 1.35 for temporary condition).

A cracked section of grout should be considered when determining resisting moment of the composite section. Particular attention should be paid to the water / cement ratio to assure grout quality. Calculations indicate a higher resisting moment can typically be achieved by increasing drill hole diameter rather than increasing the diameter of the high strength steel bar reinforcement.

Calculations and observations indicate that grout columns that are 152 mm (6 in) in diameter and are spaced 0.61 m (2 ft) on center exhibit approximately 22 kN to 44 kN (5 kips to 10 kips) of ultimate horizontal resistance during soil nail earth retention construction for heights up to 2.3 m (7 feet).

Vertical Resistance

Conventional bearing capacity theory is typically utilized to determine the bearing capacity of a reinforced block of ground such as a soil nail earth retention system. Further, charts are often utilized to assess heave potential at the bottom of an excavation and check against basal heave. Literature sources state that as long as soil nail earth retention is not constructed in soft soils, bearing capacity failure mode is not critical for most soil nail projects (FHWA 1996, FHWA 2003).

The authors recognize numerous soil nail earth retention projects are successfully completed throughout the country each year utilizing traditional bearing capacity analysis. However, from observation of construction of several soil nail earth retention projects, the authors believe bearing capacity may have larger importance to the performance (vertical and horizontal displacements) of soil nail earth retention than indicated in the literature.

In order to perform bearing capacity analysis, at least one geometric parameter (such as the length to which the bearing load is applied) has to be determined. Typically, the length of soil nail located in the reinforced block of ground is utilized as the bearing length. It is usual for the length of soil nail to be approximately equal to the completed exposed height. In this case, sufficient length will exist to distribute the load resulting in a greater value than the Factor of Safety determined in regard to bearing capacity. When bearing capacity analysis is performed for soil nail earth retention systems, the final or completed condition is typically utilized to check bearing capacity.

From observation of construction and analysis of field data, the authors believe the most critical bearing load is only applied to the front portion of the wall during excavation, approximately one-sixth or less of the reinforcement length. In addition, the authors believe a triangular bearing load distribution (loaded side of the triangle at the face) occurs in lieu of a rectangular distribution. Further research and monitoring are needed to verify the previous statements.

During construction of a typical soil nail system, material is removed from under the previously completed lift. This directly affects the bearing resistance of the system and is, in the authors' opinion, the most critical stage of construction for soil nail walls. The temporary excavated condition below the previously installed lift presents a greater risk of bearing failure than the final or completed soil nail condition. Further, if three or more rows of soil nails exist above a temporary exposed excavation, an additional vertical load is imposed on the bearing stratum.

The use of grout columns in soil nail earth retention provides vertical resistance during excavation and increases the bearing capacity Factor of Safety. The grout columns perform similar to a deep foundation system as the vertical load is transferred from the excavated lift to deeper, more competent soil. The grout columns vertically support the upper portions of the soil nail system and allow construction of the lower soil nail system. Figure 8 illustrates the vertical forces applied to the bearing stratum during soil nail construction without and with grout columns.

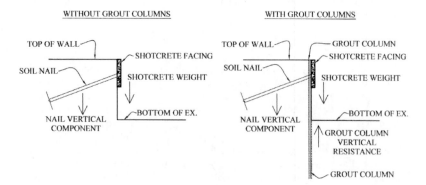

FIG. 8. Vertical Forces With and Without Grout Columns during Soil Nail Construction

As shown in Figure 8, the primary contributors to vertical load at the face of soil nail wall are the shotcrete weight and the vertical component of the soil nail due to its orientation. For typical soil nail earth retention, the shotcrete weight is resisted by the shear friction between the earth and shotcrete or the adhesion of shotcrete to the earth. The soil nail vertical component is resisted by bending stiffness of the soil nail and the interaction between the soil nail and the supporting underlying soil. As soil nail construction proceeds, the vertical load increases, inducing higher loads at the bottom of the excavation.

The use of grout columns located immediately behind the shotcrete face provides resistance to both the shotcrete weight and the soil nail vertical component. Skin friction of the grout columns below the temporary construction benches provides resistance to the imposed vertical loads.

BENEFITS OF GROUT COLUMNS IN EARTH RETENTION

Deeper First Cut to Start Under Utilities

Real estate is at a premium in many Colorado mountain towns. Consequently, developers need to utilize as much of a site as possible to ensure financial feasibility of the project. This often means the proposed structure is just a few feet away from an existing street where most underground utilities are located. Grout columns may allow the first row of soil nails to be installed deeper below existing grade, thus avoiding most utilities. Avoiding existing utilities can aid in obtaining easements and increasing productivity as individual soil nails will not have to be adjusted to ensure an existing utility is avoided.

Pre-Split Boulders

Boulders can vary in size from 0.3 m to over 1 m (1 ft to several feet) in diameter. Removal of boulders can result in damage to previously placed shotcrete, loss of soil behind the shotcrete (jeopardizing stability of the earth retention system), and/or significant overrun of shotcrete (where shotcrete is used to fill the void created by removing the boulder). Boulders encountered during grout column installation are drilled through, creating a line of weakness in the boulder (See Figure 9). When the boulder is encountered during excavation, the excavator utilizes a hydraulic hoe ram to split the boulder along the weakened plane; this results in little to no disruption to the earth retention construction.

FIG. 9. Soil Arching between Grout Columns and Pre-Spitting Boulders

Clean Face for Excavator

Maintaining a vertical face while excavating can be difficult in fine-grain soils and nearly impossible in soils containing cobbles and boulders. Grout columns provide a defined interface for the excavator to follow, increasing productivity and minimizing shotcrete overrun due to over-excavation. Grout columns reduce the likelihood that soil loss will occur.

Temporary Support between Rows

As referenced earlier, cohesionless soils tend to ravel when exposed and "run" from behind previously placed shotcrete when proceeding to the next excavation level. The loss of retained soil produces a reduction in friction at the interface between the shotcrete and the retained soil, which can result in vertical movement of the shotcrete facing. This vertical movement puts the soil nails into bending, which can jeopardize the stability of the earth retention system. Refer to earlier Figure 8.

Less Vertical and Horizontal Displacement

Published data indicate soil nail earth retention wall movement typically ranges from 0.002H (where H is the exposed height of the wall) to 0.004H (FHWA 1996, FHWA 2003). During typical soil nail construction, it is inevitable that downward movement of the face occurs during excavation due to vertical forces. This downward movement causes rotation at the top of the soil nail wall, resulting in horizontal movement of the soil nail system.

Grout columns resist the vertical forces, thus decreasing horizontal movement. Monitoring has shown that incorporating grout columns significantly reduces wall movement. Minimizing wall movement is crucial when movement tolerances have to be maintained.

Predictable Shotcrete Overrun

Incorporating grout columns in soil nail wall construction minimizes overrun due to over-excavation and filling of voids associated with removal of cobbles and boulders and the raveling of soil from behind previously placed shotcrete. Minimizing overrun:

- Allows estimators to more accurately bid earth retention projects;
- Shortens earth retention construction schedules as time allotted to fill voids and areas of over-excavation can be eliminated;
- Reduce project costs, as shotcrete can cost as much as several hundred dollars per cubic yard in Colorado mountain communities.

CASE STUDY OF SOIL NAIL CONSTRUCTION IN VAIL, COLORADO INCORPORATING GROUT COLUMNS

A recent project in Vail, Colorado demonstrates the effectiveness of grout columns in reducing soil nail earth retention wall movements. The temporary soil nail earth retention system was 4,181 m^2 (45,000 ft^2) with heights ranging from 6.1 to 16.8 meters (20 to 55 feet). Grout columns were spaced 0.61m (2 ft) on center and consisted of a 152 mm (6 in) diameter hole with a #8 Grade 520 MPa (75 kips/in^2) steel bar in the center of the hole and a 28-day compressive strength grout of 28 MPa (4,000 lb/in^2). The first row of soil nails was installed 2.1 m (7 ft) below existing grade. This required an initial cut of 1.5 m (5 ft) to install the shotcrete and reinforcing, then another cut of 1.5 m (5 ft) to install the row of soil nails. The next row of soil nails was installed 3.1 m (10 ft) below existing grade, while subsequent rows of soil nails were spaced 1.5 m (5 ft) vertically. Installing the first row of soil nails 2.1 m (7 ft) below existing grade avoided a water line, a storm sewer, and a communication utility line as shown in Figure 10.

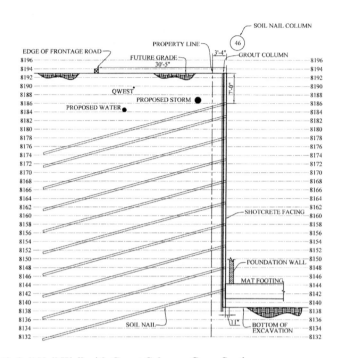

FIG. 10. Soil Nail Wall with Grout Columns Cross Section

The site soils were comprised primarily of sand, gravel, and cobbles with occasional boulders. The site was de-watered prior to construction as the static groundwater elevation was above the proposed footing elevation.

An inclinometer was installed approximately 0.61 m (2 ft) behind the face of the soil nail wall. Inclinometer readings (by CTL Thompson, Inc.) were conducted monthly from January, 2008 (just prior to beginning mass excavation) through June, 2008 when the inclinometer was accidently destroyed by the general contractor. The bottom of the excavation was attained in mid-May 2008 at which time construction started on the proposed structure mat footing and foundation wall. See Figure 11 for progressive inclinometer readings during soil nail construction.

Estimated wall movements ranged from 0.002H to 0.005H as shown in Figure 12. The cantilever portion of the soil nail wall produced maximum horizontal movement of 31.5 mm (1.24 in), which was less than the minimum estimated movement of 33.5 mm (1.32 in) at the time of inclinometer destruction.

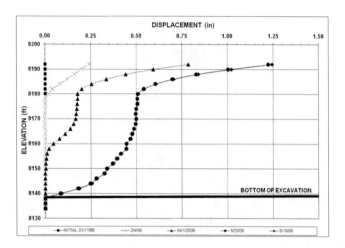

FIG. 11. Progressive Inclinometer Readings during Soil Nail Construction

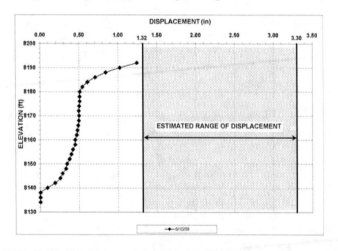

FIG. 12. Final Inclinometer Reading compared to Estimated Range

CONCLUSIONS

In conclusion, the authors believe that incorporating grout columns permits the use of soil nail construction in Colorado mountain regions where soil nail earth retention systems would otherwise not be feasible. By utilizing grout columns, it was found that the performance of soil nail earth retention was improved.

Grout columns enhance constructability by lowering the first row of soil nails to avoid utilities, ease excavation difficulty by pre-splitting boulders and establishing a defined boundary of excavation. Grout columns significantly reduce the loss of soil from behind previously placed shotcrete and provide horizontal and vertical support during excavation. The additional horizontal and vertical resistance provided by the grout columns reduces overall movement of the soil nail earth retention system thus increasing stability.

ACKNOWLEDGMENTS

The authors appreciate the insight of Larry Coggins into all aspects of earth retention and whose support is greatly appreciated. Several employees at Coggins and Sons, Inc. have assisted in the construction of soil nail walls incorporating grout columns and their input has been valuable. Tonya Hart, whose figure generation enhanced the aesthetics of the paper. Finally, Eric Martz!

REFERENCES

Abramson, L.W., Lee, T.S., Sharma, S., Boyce, G.M., "Slope Stability and Stabilization Methods." John Wiley and Sons, Inc., New York, 1996

FHWA, 2003, "Geotechnical Circular No. 7. Soil Nail Walls", Publication FHWA-IF-03-017, U.S. Department of Transportation, Federal Highway Administration, Washington, D.C.

FHWA, 1996, "Manual for Design and Construction Monitoring of Soil Nail Walls", FHWA-SA-96-096R. U.S. Department of Transportation, Federal Highway Administration, Washington, D.C.

Ito, T., Matsui, T., 1975, "Methods to Estimate Lateral Force Acting on Stabilizing Piles," Soils and Foundations, Vol. 15, No. 4, pg. 43-59

Joshi, B., September 2003, "Behavior of Calculated Nail Head Strength in Soil-Nailed Structures." Journal of Geotechnical and Geoenvironmental Engineering, ASCE, pg. 819-828

Kutschke, W.G., Tarquinio, F.S., Peterson, W.K., 2007, "Practical Soil Nail Wall Design and Constructability Issues." 32nd DFI Annual Conference, pg. 83-91

Pearlman, S.L., Campbell, B.D., Withiam, J.L., 1992, "Slope Stabilization Using In-Situ Earth Reinforcements." Geotechnical Special Publication No. 31, "Stability and Performance of Slopes and Embankments – II", June 29 - July 1, 1992, pg. 1333-1348.

Sheahan, T.C, Oral T., June 1998, "Monitoring and Failure Analysis of a Soil Nail Clay Wall." Cooperative Research Project Sponsored by ADSC and FHWA

Stocker, M.F., Riedinger, G., "The Bearing Behaviour of Nailed Retaining Structures." Geotechnical Special Publication No. 25, "Design and Performance of Earth Retaining Structures", June 18-21, 1990, pg. 612-628

Wolosick, J.R., October 1988, "Soil Nailing – A Nashville Fault Zone." Deep Foundations Institute, 13th Annual Meeting

Micropiles for Slope Stabilization

William K. Howe[1], S.M. ASCE

[1]MSCE Graduate Student, Dept. of Civil Engineering, University of Colorado Denver, Denver, CO; howe2020@hotmail.com

ABSTRACT: Micropiles have been primarily used as foundation support elements to resist static and dynamic loadings, and to a lesser extent, to provide reinforcements to the stabilization of slopes and excavations. The purpose of this paper is to determine fundamental design guidance for using micropiles for the in-situ stabilization of slope failures by performing slope stability analyses on case studies using limit equilibrium software and finite element software.

INTRODUCTION

A micropile is a small diameter drilled and grouted pile that is typically reinforced. The diameter is usually less than twelve inches and this type of pile would be considered a non-displacement pile. Micropiles can be installed at any angle, where access is restrictive, and in virtually all soil types and ground conditions.

Micropiles are used for slope stabilization to provide the necessary restraining forces to structurally support the slope. Battered, and possibly vertical, micropiles are installed through the unstable slope to a designed depth below the failure surface to establish a system similarly to Figure 1. In doing this, the micropiles provide axial, shear, and bending resistance. Most importantly, they help resist the shear forces that develop along the failure surface.

Slope stability analyses were completed for three case studies using micropiles for slope stabilization. Two of the case studies were based on actual slide events and the third was hypothetical. All case studies had some significant slope movements prior to the micropiles being designed and constructed. The analysis evaluated the factor of safety of the existing slope by performing stability analysis of the existing slope, and adjusting soil strength parameters until the factor of safety (FS) equals 1.0 for back-analysis of soil strength parameters. The optimum location of the micropile was then established and a simple method of doing so was outlined.

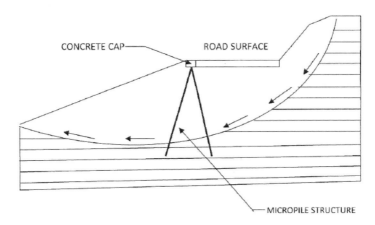

FIG. 1. Micropile System

An analysis was completed in which the upslope and down slope batter of the micropiles was varied to determine the optimum angles for the piles to be placed, and to attempt to determine a relationship between batter and factor of safety.

Lastly, the outcomes of this study were evaluated and compared to the published design for each case study.

SOFTWARE

Two software programs were used for the slope stability analysis described in the following sections. The finite element software SSI2D was used for the majority of the analysis and the limit equilibrium software STABR was used primarily as a check to the finite element software. The SSI2D software is provided by SSISoft Software Company. SSI2D software analyzes slope stability and seepage stability by two-dimensional model of the finite element method. Behavior of the soil can be described by the model as:

1- linear-elastic or

2-elastic plastic (Mohr-Coulomb)

SSI2D uses 6 node triangular and rectangular elements for soil modeling and the finite element meshing is automatic. For slope stability problems, SSI2D calculates the factor of safety.

The SSI2D program is capable of analyzing irregular slope profiles and multiple soil layers with different properties. The program does not take into account soil-structure interface so the micropile is basically modeled as a zone of high strength elastic material. The micropile can be modeled, however, at any width and at any angle or position which allows for the analysis of battered and very narrow piles.

The limit equilibrium software STABR was published in 1985 by J.M. Duncan and Kai Sin Wong of Virginia Tech University. The program calculates the factors of safety, or searches for the circular failure surface having the minimum factor of safety, using the Ordinary Method of Slices or Bishop's Modified Method.

The STABR program is capable of analyzing irregular slope profiles, tension cracks, soil layers with different properties and nonuniform thickness. Complicated pore pressure patterns and irregular variations of undrained strength with depth can also be analyzed. The slope geometry is defined by up to 20 vertical sections and the elevation of the different material boundaries at each vertical section. This input format works well for the incorporation of vertical zones of high strength material used to represent micropiles within the slope.

CASE STUDIES

The first case study was taken from the Federal Highway Administration Reference Manual No. FHWA NHI-05-039 December 2005 titled Micropile Design and Construction being described in this paper as the FHWA Design Example (U.S. Department of Transportation Federal Highway Administration, 2005). This is a hypothetical case study that was used as an example of how to design micropiles for the use in slope stabilization. The design example provided very good detail, procedures, and results that could be easily compared.

The second case study was the Littleville landslide on US Route 43 in Alabama. The slope is composed of sandstone and shale cut from adjacent hillsides. The underlying stratum is shale with some weathering along the top surface (Dan Brown, 1995). This is believed to be the failure surface. The as-constructed micropiles were 0.114 meter diameter steel casing with a wall thickness of 8 mm. The lengths were 7 meters and were battered 30 degrees from the vertical upslope and downslope. 432 micropiles and 44 ground anchors were constructed. The US 43 case study used ground anchors as well as micropiles for the slope stabilization design. However, the analysis performed in this paper used only micropiles to model the effect on slope stability.

The third case study was a root-pile wall that was constructed for the Pennsylvania Department of Transportation to correct a landslide near Monessen, Pennsylvania. The fill consisted of silty clays and clayey silts (AASHTO A-6 and A-7) intermixed with rock fragments, cinders, and building materials. The bedrock is hard sandstone to red shale with minor limestone interbeds. Groundwater is perched on the bedrock surface (Jovino, 1908). The failure surface was also assumed to run along the top of the bedrock layer. The analysis performed for this paper was only the use of micropiles as a non reticulated structure.

SLOPE STABILITY ANALYSIS

Evaluate Factor of Safety of Existing Slope

Slope stability analysis is used to evaluate the factor of safety and the estimated soil parameters. Presumably the slope stability factor of safety is approximately 1.0 if the slope has undergone significant movement. The technique used to adjust the soil strength parameters is termed back-analysis (U.S. Department of Transportation Federal Highway Administration, 2005). Back-analysis involves the following steps:

1) Estimate soil parameters of existing slope
2) Perform slope stability analysis
3) Adjust parameters until FS = 1.0

The location of the critical failure surface should be considered during the back-analysis. Not only should the factor of safety be approximately 1.0, but the failure surface should also match reasonably to that observed in the field.

TABLE 1. Comparison of back-analysis of soil strength parameters

SOFTWARE OR METHOD USED	FHWA DESIGN EXAMPLE			US 43 CASE STUDY			PENNSYLVANIA CASE STUDY		
	SLIDE	SSI2D	STABR	STABL5	SSI2D	STABR	MORGENSTERN-PRICE	SSI2D	STABR
c^a	15	15	15	NOT PROVIDED	0	0	4.79	5	5
ϕ^b	11	11	12	NOT PROVIDED	19	20	17	17.5	17.5

[a] c represents cohesion in units of kN/m^2

[b] ϕ represents angle of internal friction in degrees from horizontal

A slope stability analysis using SSI2D was completed for the three case studies and the factor of safety using the trial soil parameters was determined. The cohesion and internal friction angle of the fill and residual soils were modified until a factor of safety of nearly 1.0 was obtained. The analysis showed that the SSI2D and STABR programs produce very similar results as far as the factor of safety of an existing slope and the soil strength parameters determined from back-analysis. These compared very well to the two case studies that used different software and provided the soil strength parameters they produced. These two case studies were the FHWA design example and the Monessen, Pennsylvania case study. A comparison is made in Table 1. The US 43 case study did not provide the details of the soil strength parameters determined from back-analysis.

The failure surface was then checked and found to be reasonably consistent with the actual observed failure surface. The displacement contour is used in this analysis as the primary tool to match the software's failure surface to the observed failure surface location. The results are shown in the displacement contours of Figures 2, 3, and 4.

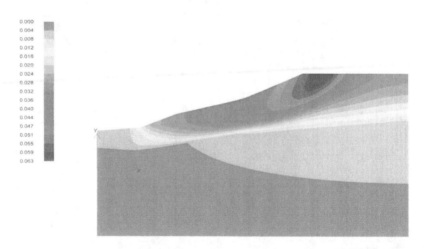

FIG. 2. Total displacement contour in meters for FHWA design example using modified strength parameters (FS = 1.04)

FIG. 3. Total displacement contour in meters for US 43 case study using modified strength parameters (FS = 1.04)

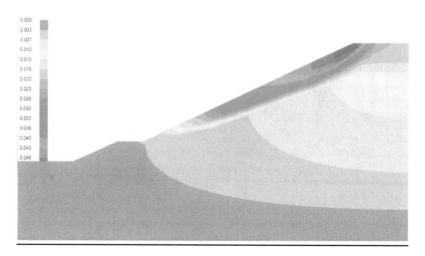

FIG. 4. Total displacement contour in meters for Monessen, Pennsylvania case study using modified strength parameters (FS = 0.97)

Determine Location of Micropile Structure

Analysis was performed to determine the optimum location within the slope for the micropile. Several trials were performed for each case study using both the SSI2D and STABR programs. A single vertical pile of width 0.5 meters was used in both the SSI2D analysis and STABR analysis. This was done in the SSI2D program in order to minimize software run time and come up with a reasonable procedure to determine the location. The run time for a 0.229 m battered micropile pair is on average 6 hours. There is also more time needed to draw in the geometry of the smaller, angled, micropiles. The run time for a 0.5 m vertical pile is on average 10 minutes. The results obtained from the 0.5 m vertical pile are the same as two 0.229 m piles in terms of the location within the slope that provides the largest factor of safety, or is the most substantial in terms of preventing failure of the slope.

The micropiles were varied within the failure surface, as determined from the displacement contours above, approximately every 5 meters. The factor of safety versus a normalized distance from the toe of slope was plotted to show the optimum location for the micropile. The results from the STABR program and the SSI2D program were plotted on the same axis to compare the results.

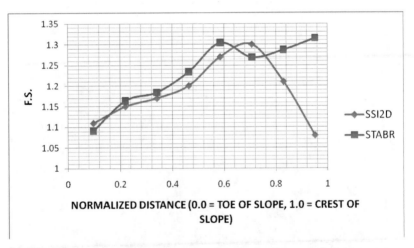

FIG. 5. Relationship between FS and normalized distance from toe of slope for FHWA design example

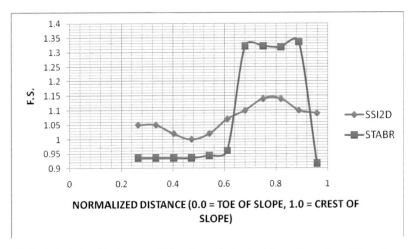

FIG. 6. Relationship between FS and normalized distance from toe of slope for US 43 case study

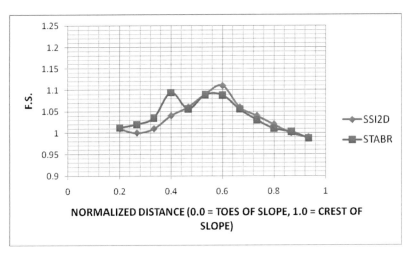

FIG. 7. Relationship between FS and normalized distance from toe of slope for Monessen, Pennsylvania case study

Determine Batter of Micropile Up and Down Slope

The upslope and down slope batter of the micropiles was varied to determine the optimum angles for the piles to be placed, and to attempt to determine a relationship between batter and factor of safety. A graphical representation of the results did not provide a clear relationship, therefore, the results are provided in tabular form as shown in Table 2. Of the two programs only the SSI2D program is capable of doing a battered analysis.

Due to the amount of run time required to do such an analysis only the FHWA design example was evaluated for the effect of batter. The angles used for the upslope and down slope batter of the FHWA design example were then used for the remaining two case studies. These angles are similar to those used in most case studies encountered for palpable reasons such as construction costs, and that a down slope battered pile can provide more capacity by developing axial forces rather than bending forces.

The optimum batter angles from the FHWA design example were determined to be 30° down slope and 10° upslope. As can be seen from Table 2, the maximum factor of safety was encountered when upslope angle was 10° and the down slope angle was 30° or larger. To limit the study to within practical construction constraints and because the larger down slope angles would be more costly, in other words longer piles, the angle of 30° was selected.

Note that the SSI2D program would not mesh a pair of 0.229 m micropile at a combined angle less than 40° (10° up and 30° down).

TABLE 2. Micropile batter versus factor of safety

F.S.	DOWN SLOPE BATTER (degrees from vertical)	UP SLOPE BATTER (degrees from vertical)
1.35	10	30
1.33	10	40
1.31	10	50
1.36	30	10
1.29	30	15
1.28	30	20
1.33	30	30
1.33	30	40
1.32	30	50
1.36	40	10
1.36	50	10
1.36	60	10

RESULTS

In comparing and evaluating the outcomes of this study to the published design for each case study, they were found to be similar. The FHWA final design evaluation, in comparison, is almost identical to the FHWA design. The soil parameters obtained from back analysis were identical, with the fill and residual soil both having a φ = 11° and c = 15 kN/m². The pile cap locations were almost identical at 57% up the slope from the toe, and with a factor of safety of 1.30 obtained in their slope stability analysis versus 1.36 obtained here. The published design micropiles were battered at 3° upslope and 20° down slope with reasoning being simply that "...the pair be constructed as close to each other as possible with just large enough space to permit ease of construction" (U.S. Department of Transportation Federal Highway Administration, 2005). In contrast, the micropiles evaluated in this paper were battered at 10° upslope and 30° down slope.

There are some differences in comparison to the US 43 design. The main difference being the actual design consisted of two battered micropiles and a ground anchor coming from the same pile cap. This paper focuses only on using micropiles in slope stabilization so this difference is unavoidable. The soil parameters obtained from back analysis could not be compared as there was no specific information given in the literature. The pile cap location was approximately 67% up the slope from the toe for the actual design versus at 79% in this design. The final factors of safety could not be compared as this information was also not given.

Another difference is that the micropiles of the actual design were both battered upslope and down slope at an angle of 30° from the horizontal (versus 10° and 30°) (Dan Brown, 1995). The micropiles consisted of a steel casing and grout composite pile, however, the micropiles were only 152 mm o.d. versus 229 mm that was used in this analysis.

There are differences in comparison to the Pennsylvania design case study. The main difference being the actual design consisted of a root-pile wall. The design procedure is very simplistic and similar to designing a gravity retaining wall. Another difference is the micropiles used a No. 9 reinforcing steel bar versus steel casing and grout composite pile. Additionally, the micropiles were only 127 mm o.d. versus 229 mm that was used in this analysis (Jovino, 1908).

The resulting conclusion is that micropiles alone could not be used to stabilize the slope in this third case study. A removal and replacement of some of the soil, as was done in the actual case, could provide enough stability for the micropile to be used.

CONCLUSIONS

A slope stability analysis was performed for three case studies using micropiles for slope stabilization. All case studies had some significant slope movements prior to the micropiles being designed and constructed. The analysis consisted primarily of the following:

- Evaluate the factor of safety of the existing slope by
  - o Performing stability analysis of the existing slope
  - o Adjusting soil strength parameters until FS=1.0 for back-analysis of soil strength parameters
- Determine the optimum location of the micropile and establish a simple method of doing so
- Determine the batter of the micropile

Many trials were performed to determine the optimum location of the micropile within a slope for stabilization purposes. The results show that the optimum location is near the middle of the circular failure surface, or more importantly, near the point of tangency of the failure surface to the bedrock or "stiff" soil plane. This is the result for all three case studies evaluated and is not expected to differ as long the slope consists of a failure plane along a bedrock/soil interface.

An effort was also made to provide a plot for the generalized optimum location for the micropile within the slope that could possibly be used for different slope sizes or profiles. Figures 8 and 9 show the normalized distance from the toe of slope for all three case studies. Both figures show the optimum location is near the middle of the slope, but more specifically, the optimum location can usually be said to be within the third-quarter of the slope, or from 50% to 75% up the slope from the toe. When compared to studies utilizing conventional large pile systems, these findings suggest, as S. Hassiotis et al. concluded, that the upper middle part of a slope is the optimum location (S. Hassiotis, 1997). This is also in line with the Laudeman thesis that concluded the optimum location is within the upper one third of the slope (Laudeman, 2002).

Methodologies have been proposed to support conclusions of other studies in regard to where a conventional pile should be located along the length of the slope. S. Hassiotis et al. explained that for a maximum factor of safety, the piles must be placed in the upper middle part of the slope. Generally, they must be located closer to the top of steeper slopes than shallower slopes (S. Hassiotis, 1997). However, in this study, as well as most others, very little data and not a significant amount of trials appear to be performed to determine these conclusions.

The data presented in this paper provides a simple process to find the optimum location for micropile. This process consists of first obtaining and inputting the actual slope geometry and soil profile, and then determining and field verifying the location of the critical failure surface by slope stability analysis. The optimum location will then most likely be located near the middle of the slope, or more importantly will be located near the point of tangency of a circular failure surface. The engineer should evaluate each slope for the appropriate design.

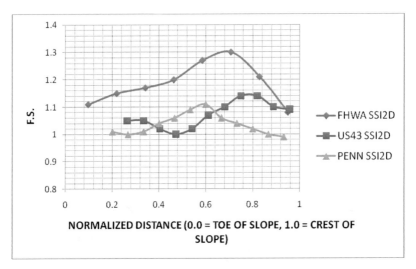

FIG. 8. Relationship between FS and normalized distance from toe of slope using SSI2D software

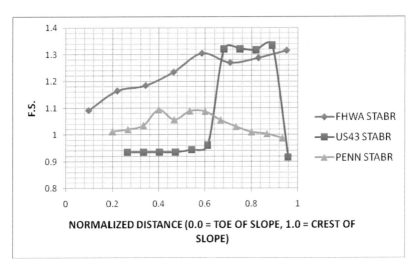

FIG. 9. Relationship between FS and normalized distance from toe of slope using STABR software

ACKNOWLEDGEMENTS

The author appreciates the support and contributions made by Dr. N.Y. Chang and the University of Colorado Denver.

REFERENCE

Dan Brown, a. J. (1995). Instrumentation and Mearurements of a Type-A Insert Wall. *ASCE Special Publication No. 50 Upgrading Foundation and Repair* , 27-41.

Jovino, U. D. (1980). Construction of a Root-Pile Wall at Monessen, Pennsylvania. *Transportation Research Record* , 13-21.

Laudeman, S. (2002). *Finite Element Analysis of Slope Stabilization Using Piles.* Denver, CO: University of Colorado Denver.

S. Hassiotis, J. C. (1997). Design Method for Stabilization of Slopes with Piles. *Journal of Geotechnical and Geoenvironmental Engineering* .

U.S. Department of Transportation Federal Highway Administration. (2005). *Micropile Design and Construction.* Washington, D.C.: National Highway Institute.

Probabilistic and Deterministic Slope Stability Analysis
by Random Finite Elements

Giorgia F. deWolfe[1] S.M. ASCE, D.V. Griffiths[2] F. ASCE, Jinsong Huang[3] M. ASCE

[1]Civil Engineer, PhD, Bureau of Reclamation, Denver Federal Center, Denver, Colorado, U.S.A; email: gdewolfe@usbr.gov
[2]Division of Engineering, Colorado School of Mines, Golden, Colorado, 80401-1887, U.S.A; email: dvgriffiths@mines.edu
[3] Division of Engineering, Colorado School of Mines, Golden, Colorado, 80401-1887, U.S.A; email: jhuang@mines.edu

ABSTRACT: Program PES (Probabilistic Engineered Slopes) provides a repeatable methodology allowing the user to perform a slope stability analysis on a one-sided and two-sided sloping structure using a deterministic or probabilistic approach. Program PES, in contrast with other deterministic or probabilistic classical slope stability methodologies, is cable of seeking out the critical failure surface without assigning a pre-defined failure surface geometry. The probabilistic approach of program PES applies the Random Finite Element Method (RFEM) by Griffiths and Fenton (1993) taking into account the soil spatial variability and allowing the use of different random fields to characterize the spatial variation of any material type. The methodology is compared against the probabilistic approach proposed by the program SLOPE/W version 7.14 (Geostudio Group, 2007), and demonstrates its potential for predicting probability of failure (p_f) in non-homogeneous soil structures characterized by phreatic conditions and potential post-earthquake liquefiable conditions. The p_f results obtained by program PES have proved that underestimating the influence that the soil material variability has on the computation of p_f will lead to lower results of probability and underestimate of the risk of slope instability. Program PES capabilities could be used by the engineering practice to prioritize intervention activities within a risk context.

INTRODUCTION

Stability analyses are routinely performed in order to assess the equilibrium conditions of natural and manmade slopes. The analysis technique chosen depends on both site conditions and the potential mode of failure, with careful consideration being given to the varying strengths, weaknesses and limitations inherent in each methodology.

The motivation driving this study is closely related to the assessment and mitigation of the hazards caused by the instability processes and the important role that stability analysis of slopes plays in civil engineering applications and design.

For many years the nature of geotechnical slope stability analysis has been predominantly deterministic, whether performed using design charts or computers.

It is inherent in this type of approach that the parameters characterizing the soil materials such as friction angle, cohesion, Young's modulus, Poisson's ratio, unit weight and ground water are also treated as deterministic. Intuitively, it can be recognized that, where there are materials more homogeneous than others in nature, there are no perfectly homogeneous natural materials. The deterministic approach, which does not allow any variation in the soil materials properties, clearly introduces a high level of approximation to the analysis and characterization of slope stability. The level of approximation can only be reduced if the natural variation of soil is taken into account, allowing the soil to be characterized by a range of values for each parameter instead of a single value.

Soil properties measurements are usually taken over a finite volume, which represents a local average of the property with respect to the overall size of the site domain. For this reason, the Local Average Subdivision (LAS) method (Fenton and Vanmarcke, 1990) has been used to generate the random fields in all the investigations presented in this work. The random field model provides a useful tool for the generation of spatially variable soil properties. A random field is characterized by sets of soil property values, which are randomly generated around their mean value, and are mapped onto the finite element mesh creating a 2D model of variable soil. Each set of property values (e.g. cohesion and friction) characterizes an element within the domain analyzed. The Monte-Carlo method is lastly applied to this model performing multiple random field realizations. The number of simulation that give a Factor of Safety (FS)<1 divided by the total number of simulations represents the probability of failure.

In the computation of slope stability and probability of failure certainly there are many sources of uncertainty, in addition to those related to soil variability. In current engineering practice, most slope stability analyses following a deterministic approach or characterized by a 1D model, do not account for soil variability. The current work will show that accounting for the influence of soil variability, varying the soil strength parameters and using a 2D model, leads to more conservative probability of failure results compared to those computed using classical approaches to geotechnical problems.

PROGRAM PES CHARACTERISTICS

Deterministic Theory

Program PES (Probabilistic Engineered Slopes) coded in FORTRAN.95 allows the user to perform a slope stability analysis on a one-sided and two-sided sloping structure using a deterministic or probabilistic approach. A brief description of program PES methodology is given below. For more detailed information on the elastic- visco-plastic and the strength reduction algorithms used in this study the reader is referred to Griffiths and Lane (1999) and Smith and Griffiths (2004).

As a first step program PES reads the geometry input parameters from the input data file generating a finite element mesh of the problem. Subsequently the soils properties recorded in the input data file are assigned to the relative embankment and foundation deterministic mesh regions and random fields.

The program can allow the analysis of a liquefiable layer ether in the foundation or in the embankment as well as the partition of a homogeneous embankment into two materials. Clearly these more complicated components are highly dependent on the problem analyzed and require modifications of the main program code each time a different problem is selected.

After the information from the input data files are read Program PES computes the elastic stress-strain matrix, the shape function at the integrating points, the analytical version of the stiffness matrix for an 8-node quadrilateral element, and the lower triangular global matrix kv. Then the program generates the additional loading due to free-standing water outside of the slope, as well as a pore pressure within the slope. The water load is equal to the summation of gravity load and pore pressure load, and is computed before being added to the total load already computed. The program allows for the analysis of submerged slopes as well as slopes characterized by a specific water table which can vary in elevation throughout the mesh.

Subsequently program PES computes the strength reduction factor and then performs a check on whether or not the yield is violated according to the failure criterion. The theory coded in this section of the program is described in more detail in the following paragraphs.

Program PES models a 2D plane strain analysis of elastic-perfectly plastic soils with a Mohr-Coulomb failure criterion using 8-node quadrilateral elements with reduced integration (4 Gaussian-points per element) in the gravity load generation, the stiffness matrix generation and the stress redistribution phases of the algorithm. From the literature, conical failure criteria are the most appropriate to describe the behavior of soils with both frictional and cohesive components, and the Mohr-Coulomb criterion is known as the best of this group of failure criteria. Therefore the program uses the Mohr-Coulomb criterion as failure mechanism in all cases. In terms of principal stresses and assuming a compression-negative sign convention, the Mohr-Coulomb criterion can be written as shown in Eq. 1

$$F_{mc} = \frac{\sigma_1' + \sigma_3'}{2} \sin \phi' - \frac{\sigma_1' - \sigma_3'}{2} - c' \cos \phi' \tag{1}$$

where σ_1' and σ_3' are the major and minor principal effective stresses.

In cases where the soil is characterized by a frictionless component (undrained clays) the Mohr-Coulomb criteria can be simplified into the Tresca criterion substituting $\phi = 0$ in Eq. 1 and obtaining Eq. 2,

$$F_t = \frac{\overline{\sigma}(\cos \theta)}{\sqrt{3}} - c_u \tag{2}$$

The failure function F for both criteria can be interpreted as follows:

F<0 stresses inside failure envelope (elastic)
F=0 stresses on failure envelope (yielding)
F>0 stresses outside failure envelope (yielding and must be redistributed)

The soil is initially assumed to be elastic and the model generates normal and shear stresses at all Gauss-points within the mesh. These stresses are then compared with the Mohr-Coulomb failure criterion.

The elastic parameter E' and υ' refer to Young's modulus and Poisson's ratio of the soil, respectively. If a value of Poisson's ratio is assumed (typical drained values lie in the range 0.2 $<\upsilon'<$0.3), the value of Young's modulus can be related to the compressibility of the soil as measured in a 1D oedometer (e.g. Lambe and Whitman 1969) as shown in Eq. 3,

$$E' = \frac{(1+\upsilon')(1-2\upsilon')}{m_\upsilon(1-\upsilon')} \tag{3}$$

where m_v is the coefficient of volume compressibility.

In this study the parameters E' and υ' have the values of (E'=10^5 kN/m^2 and υ'=0.3) respectively. The total unit weight γ assigned to the soil is proportional to the nodal self-weight loads generated by gravity. The forces generated by the self weight of the soil are computed using a gravity procedure which applies a single gravity increment to the slope. The gravity load vector for a material with unit weight γ is computed at the element level as shown in Eq. 4, and subsequently accumulated from each element at the global level by integration of the shape function [N] as shown in Eq. 5,

$$gravlo^{(e)} = \gamma \int_{V^e} N^T dV^e \tag{4}$$

$$gravlo = \sum_{elemnts}^{all} \gamma \iint [N]^T dxdy \tag{5}$$

where N represents the shape functions of the element and the superscript e refers to the element number. This integral evaluates the volume of each element, multiplies by the total unit weight of the soil and distributes the net vertical force consistently to all the nodes.

Others have shown that in nonlinear analyses, the stress paths due to sequential loading versus the path followed by a single increment to an initially stress-free slope can be quite different; however the factor of safety appears unaffected when using elasto-plastic models (e.g. Borja *et al* 1989, Smith and Griffiths 2004). It is also important to remember that classical limit equilibrium methods do not account for loading sequence in their solutions.

In the program the application of gravity loading is followed by a systematic reduction in soil strength until failure occurs. This is achieved using a strength reduction factor *SRF* which is applied to the frictional and cohesive components of strength in the form of Eq. 6

$$\phi_f' = \arctan\left(\frac{\tan\phi'}{SRF}\right) \quad \text{and} \quad c_f' = \frac{c'}{SRF} \tag{6}$$

The factored soil properties ϕ_f' and c_f' are the properties actually used in each trial analysis. When slope failure occurs, as indicated by an inability of the algorithm to find an equilibrium stress field that satisfies the Mohr-Coulomb failure criterion coupled with significantly increasing nodal displacements, the factor of safety is given by Eq. 7

$$FS \approx SRF \tag{7}$$

In the literature this method is referred to as the "shear strength reduction technique" (e.g. Matsui and San 1992).

The reduction of soil strength is followed in the program by the computation of the total body load vectors. A description of generation of the body loads computed in the program can be found in deWolfe (2010) and a detailed description of the algorithm used in the program involving viscoplasticity can be found in Smith and Griffiths (2004).

After the computation of body load vectors is completed the program generates the graphic output files respectively a PostScript image of the nodal displacement vectors and a PostScript image of the deformed mesh. The PostScript plot of the displaced finite element mesh has an optional grey-scale representation of the material property random field.

Probabilistic Theory

With regard to the probabilistic analysis computed by program PES, the probability of failure can be calculated using two different approaches. When the program is asked to compute the safety factor (*FS*) for each Monte-Carlo simulation, the probability of failure is described by the proportion of Monte-Carlo simulations with *FS*<1. When the program is asked to compute the probability without determining the exact value of *FS* for each simulation, the probability of failure is described by the proportion of Monte-Carlo slope stability analyses that failed. In this case the SRF is equal to 1(no strength reduction is actually applied). In this case, "failure" was said to have occurred if, for any given realization, the algorithm (Mohr-Coulomb failure criterion) was unable to converge within 500 iterations.

The RFEM code enables a random field of shear strength values to be generated and mapped onto the finite elements mesh, taking full account of element size in the local averaging process. In a random field, the value assigned to each cell (or finite elements in this case) is itself a random variable.

The random variables can be correlated to one another by controlling the spatial correlation length and the cross correlation matrix where the degree of correlation ρ between each property can be expressed in the range of $-1 < \rho < 1$.

More generally the correlation coefficient between two random variables X and Y can be defined by Eq. 8

$$\rho_{XY} = \frac{COV[X,Y]}{\sigma_x \sigma_y} \tag{8}$$

where COV represents the covariance between the two variables X and Y and their respective standard deviations σ_x and σ_y.

Due to the isotropic approach applied throughout this work the following simplifications can be made with respect to the mean, standard deviation and the spatial correlation length: $\mu_x = \mu_y = \mu_z$, $\sigma_x = \sigma_y = \sigma_z$, and $\theta_x = \theta_y = \theta_z$.

Using an exponentially decaying (Markovian) correlation function, Eq. 8 can be rewritten as in Eq. 9 and Eq. 10

$$\rho = e^{-\frac{2\tau}{\theta_{\ln c}}} \tag{9}$$

$$\rho = \exp\left\{ -\frac{2}{\theta_{\ln c}} \sqrt{\tau_x^2 + \tau_y^2} \right\} \tag{10}$$

Where ρ is the familiar correlation coefficient, τ is the distance between two points in the random field and $\theta_{\ln c}$ represent the spatial correlation length.

The spatial correlation length (θ), also referred to in literature as the "scale of fluctuation", describes the distance over which the spatially random values will tend to be significantly correlated in the underlying Gaussian field. Mathematically θ is defined as the area under the following correlation function (e.g. Fenton and Griffiths, 2008 from Vanmarcke, 1983);

$$\theta = \int_{-\infty}^{\infty} \rho(\tau) d\tau = 2 \int_{0}^{\infty} \rho(\tau) d\tau \tag{11}$$

where τ represents the distance between two positions in the random field. A large value of θ will imply a smoothly varying field, while a small value will imply a ragged field.

Another important dimensionless statistical parameter involved in this probabilistic approach is the coefficient of variation v, which for any soil property can be defined as

$$v = \frac{\sigma}{\mu} \tag{12}$$

where σ is the standard deviation and μ the mean value of the property.

In brief, the analyses involve the application of gravity loading, and the monitoring of stresses at all the Gauss points. The program uses the Mohr-Coulomb failure criterion, which if violated, attempts to redistribute excess stresses to neighboring elements that still have reserves of strength. This is an iterative process which continues until the Mohr-Coulomb criterion and global equilibrium are satisfied at all points within the mesh under quite strict tolerances. Plastic stress redistribution is accomplished using a visco-plastic algorithm with 8-node quadrilateral elements and reduced integration in both the stiffness and stress redistribution parts of the algorithm. For a given set of input shear strength parameters (mean, standard deviation and spatial correlation length), Monte-Carlo simulations are performed until the statistics of the output quantities of interest become stable.

A more comprehensive explanation of the random finite elements method, including local averaging approach and discussion on spatial correlation length can be found in Fenton and Griffiths (2008).

PROGRAM PES APPLICATIONS

Fruitgrowers Dam Deterministic and Probabilistic Slope Stability Analyses

In this section program PES is tested in the analysis of a dam case history. Fruitgrowers Dam is located in Delta County, Colorado, 6.4 kilometers upstream from Austin, Colorado on Alfalfa Run, a tributary of the Gunnison River. The dam was constructed by the Bureau of Reclamation from 1938 to 1939 for the primary purpose of irrigation. The crest of the dam is at elevation 1674.0 meters (5493 feet). The dam has a structural height of 16.8 meters (55 feet), hydraulic height of 12.2 meters (40 feet), crest width of 7.6 meters (25 feet), and crest length of 463.3 meters (1520 feet). An aerial view of Fruitgrowers dam is shown in Figure 1.

The dam is a compacted zoned earthfill structure consisting of a wide central core protected by a riprap layer on the upstream slope and by a thin gravel shell on the downstream slope. The embankment core is composed of clay, sand and gravel, grading to gravel at the outer slopes as shown in Figure 2. A cut-off trench was excavated to impermeable material. The trench has a bottom width of 2.4 meters (8 feet) and is located 10.7 meter (35 feet) upstream of dam centerline. The surficial material beneath the dam shell upstream and downstream of the cut-off trench was stripped to remove top soil and organic material.

FIG. 1. Aerial view of Fruitgrowers dam (Photo courtesy of the BOR)

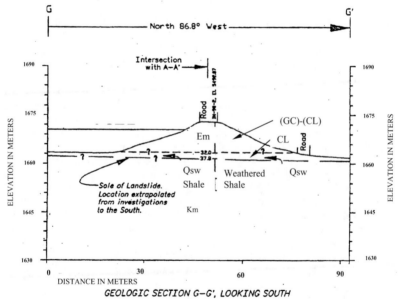

GEOLOGIC SECTION G–G', LOOKING SOUTH

Geologic label description: EM: Embankment material GC-CL: gravel with clay and sand to lean clay CL: Lean Clay Qsw: Quaternary slope wash alluvium Km: Mancos Shale Formation

FIG. 2. Cross section G-G' showing post construction actual dimensions of Fruitgrowers Dam

The case history of Fruitgrowers Dam was selected because past studies of the site conducted by the Bureau of Reclamation presented possible post-earthquake liquefiable conditions in the foundation.

A seismic hazard assessment, conducted by the Bureau of Reclamation (2003), concluded that the background earthquake sources present in the area will not likely result in a large liquefaction potential. In August 2004, to address new concerns created by the presence of silty sand material on the dam abutment, a study was conducted using data collected from five field explorations performed between 1980 and 1999.

The results of this latest study showed a low likelihood of foundation liquefaction at the dam site. According to this study, to produce the failure of the embankment a liquefied continuous lens, longer than 19.5 meters (64 feet), should be present in the foundation under the right abutment, and from the drill log data collected on each side of the embankment during the field explorations the presence of such a long continuous layer is unlikely. As shown in Figure 3, a deterministic post liquefaction FS of 1.05 was computed for the structure assuming the presence of a 18.3-meter (60-foot) long liquefiable layer.

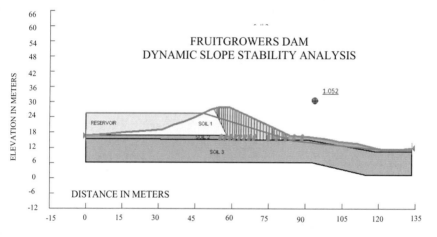

Figure 3: Deterministic post-liquefaction steady state analysis computed in 2004 using the software SLOPE/W version 7.4

From the computer program SLOPE/W version 7.4, the method of analysis used to compute this result was the Spencer method, coupled with a rigid block theory technique for the evaluation of the failure surface.

In the "Evaluation of Liquefaction and Post Earthquake Stability" conducted by the Bureau of Reclamation in August 2004, as well as in previous studies, the dam is essentially modeled as a homogeneous embankment. Similar to the study conducted in 2004 the geometry of the current model is based on cross section G-G, Figure 2 (post construction actual dimensions) and also represents a homogeneous embankment.

The phreatic condition characterizing the analysis is also adopted from the model constructed in 2004 which shows the reservoir elevation at 1672 meters (5485) (top of active conservation) with 2.44 meters (8 feet) of freeboard, and a downstream toe water elevation of 1662 meters (5453 feet), 1.22 meter (4 feet) below ground surface.

This piezometric line was developed during a study also conducted in 2004 investigating the effect of the artesian pressure on the site foundation and embankment structure (Technical Memorandum No. FW-8312-2, 2004). Figure 4 shows the piezometric line, the geometry and the major units characterizing the deterministic model created in 2004.

The model representing Fruitgrowers Dam is characterized by the following 3 soil materials.

- The embankment core is composed of clay, sand and gravel, grading to gravel at the outer slope.
- The foundation material consists of the Mancos Shale Formation (Km) and is modeled with a thickness of 11 meters (36 feet).
- The Quaternary alluvium (Qal) is characterized by recent alluvial deposits of the Alfalfa Run and is modeled with a thickness of about 1.83 meters (6 feet).

Before diving into the probabilistic analysis, initial deterministic static analyses modeling pre- and post-liquefaction conditions were conducted using program PES.

FRUITGROWERS DAM
DYNAMIC SLOPE STABILITY ANALYSIS

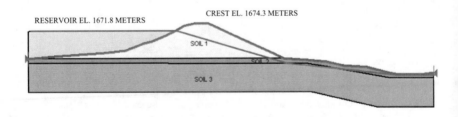

FIG. 4. Representation of the 2004 model used in the deterministic post liquefaction analysis.

The soil properties used in the 2004 slope stability analysis to characterize the embankment, foundation, and liquefiable layer are considered generally appropriate for these two deterministic analyses and are summarized in Table 1.

Table 1. Deterministic soil properties used in the Fruitgrowers Dam pre and post-liquefaction analyses

	Material	Unit weight (kg/m^3)	ϕ' (°)	c' (kPa)
Post liquefaction conditions	Embankment	2050	32	20.68
	Foundation	2082	30	0.05
	Quaternary alluvium	2082	0	14.36
Pre liquefaction conditions	Embankment	2050	32	20.68
	Foundation	2082	30	0.05
	Quaternary alluvium	2082	30	0.05

Subsequently the post liquefaction deterministic model was run using the probabilistic capability offered in program PES.

The soil properties as probabilistic variables and their statistical parameters used during the probabilistic analysis are summarized in Table 2.

Table 2. Probabilistic soil properties used in the Fruitgrowers post-liquefaction analyses

Material	μ	σ characterize by lower v	σ characterize by higher v	Distribution Type
Embankment ϕ' (°)	32	3.2	6.4	Lognormal
Embankment c' (kPa)	20.68	2.07	4.14	Lognormal
Foundation ϕ' (°)	30	6	15	Lognormal
Foundation c' (kPa)	0.05	0.009	0.02	Lognormal
Quaternary alluvium ϕ' (°)	0	0.2	0.5	Lognormal
Quaternary alluvium c' (kPa)	14.36	2.87	7.18	Lognormal

The probabilistic analysis associates one random field with the embankment, one with the foundation and the liquefiable layer is described by the foundation random field which is modified to address the new values describing the liquefiable material. In this probabilistic model only the strength parameters of friction and cohesion are analyzed in a probabilistic approach; the other parameters, dilation angle, unit weight, Young's modulus and Poisson's ratio are analyzed following a deterministic approach.

To address the level of uncertainty incorporated with the mean values describing the properties the same probabilistic model is run one time with a higher Coefficient of Variation (v) and one time with a lower v. The v values chosen represent suggested values available in the literature for similar soil material. (e.g Phoon and Kulhawy, 1999). The v values used in this analysis for all material types are summarized in Table 3

Table 3. v values characterizing Fruitgrowers probabilistic runs.

Material	lower v	higher v
Embankment ϕ' (°) and c' (kPa)	0.1	0.2
Foundation ϕ' (°) and c' (kPa)	0.2	0.5
Quaternary alluvium ϕ' (°) and c' (kPa)	0.2	0.5

The v values characterizing the probabilistic analyses were chosen evaluating suggested values available in the literature for similar soil material. (e.g Lee et al. 1983, Phoon and Kulhawy, 1999).

Another critical value in the analysis is the spatial correlation length used to determine the soil spatial variability. The set of isotropic values chosen to investigate the spatial correlation length θ for all probabilistic runs is reported in Table 4.

Table 4. Isotropic θ values characterizing Fruitgrowers spatial variation of soil.

$\theta=$	1.22 m
$\theta=$	7.62 m
$\theta=$	18.288 m
$\theta=$	30.48 m
$\theta=$	60.96 m
$\theta=$	91.44 m
$\theta=$	152.4 m

All the probabilistic analyses are run using 1000 Monte-Carlo simulations. It has been observed during this investigation that the probabilistic model representing Fruitgrowers dam associated with 1000 Monte-Carlo simulations returns a probability that can vary up to 2.7% as showed in Figure 5, which represent a repeatable computation. During all probabilistic and deterministic analyses all soil properties are considered uncorrelated between each other.

The results of the probabilistic analyses as well as the comparison with the results generated by the program Slope\W version 7.14 are described in the following section

Variability in *pf* results

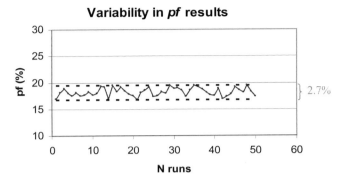

FIG. 5. Variability in p_f results using 1000 Monte-Carlo simulations. To recognize how much the p_f computed by the Fruitgrowers model could vary in a probabilistic setting the same data file was run 50 times.

Programs PES and SLOPE/W: Deterministic and Probabilistic Slope Stability Results Comparison.

The result from the deterministic pre-liquefaction model run using program PES shows a FS=1.66 (Figure 6) while the SLOPE/W result according to Spencer's Method returns a FS=1.746 (Figure 7). The deterministic post-liquefaction model computed by PES returned a value of FS=1.09 (Figure 8) when the SLOPE/W result on the same model according to Spencer's Method returned a FS=1.06 (Figure 9).

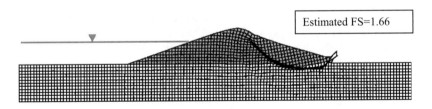

Estimated FS=1.66

FIG. 6. Displacement file showing displacement associated with the deterministic pre-liquefaction conditions at Fruitgrowers Dam.

FRUIT GROWERS DAM
STATIC SLOPE STABILITY ANALYSIS 2009
pre liquefaction material properties

Non liquefiable Km	Liquefiable weathered Km	Embankment
Unit Weight: 2082 kg/m^3	Unit Weight: 2082 kg/m^3	Unit Weight: 2050 kg/m^3
Cohesion: 0.0478 kPa	Cohesion: 0.0478 kPa	Cohesion: 20.68 kPa
Phi: 30°	Phi: 30°	Phi: 32°

FIG. 7. Graphic representation according to Spencer's Method of the SLOPE/W results describing the deterministic pre-liquefaction conditions at Fruitgrowers Dam.

The loading applied in the post liquefaction analysis are vertical gravity load only. The post liquefaction analysis results from both programs assumes the presence of a liquefiable layer, 84.12 meter (276 feet) long (16.45 meter or 54 feet downstream from the centerline of the dam), while the post liquefaction deterministic analysis computed in the 2004 obtained a FS=1.05 assuming the presence of a continuous liquefiable layer 18.29 meter (60 feet) downstream of the centerline of the dam.

In the probabilistic analysis computed by PES the deterministic variables are characterized by the same values used in the post-liquefaction analysis and the probabilistic values are described by the statistical parameters summarized in the previous section. In the probabilistic analysis computed using SLOPE/W, the failure surface associated with the FS of 1.06 (Figure 9) was chosen as the critical one to test with the probabilistic approach offered by SLOPE/W.

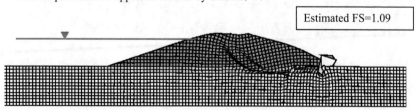

Estimated FS=1.09

FIG. 8. Representation of the displacement associated with the deterministic post-liquefaction conditions at Fruitgrowers Dam (program PES).

The soil properties statistical parameters and soil spatial variation parameters used in this analysis are the same as those used in the analysis run with program PES, and are summarized in Tables 2, 3, and 4.

FRUT GROWERS DAM
STATIC SLOPE STABILITY ANALYSIS 2009
post liquefaction material properties

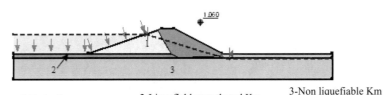

1-Embankment	2-Liquefiable weathered Km	3-Non liquefiable Km
Unit Weight: 2050 kg/m³	Unit Weight: 2082 kg/m³	Unit Weight: 2082 kg/m³
Cohesion: 20.68 kPa	Cohesion: 13.36 kPa	Cohesion: 0.0478kPa
Phi: 32°	Phi: 0°	Phi: 30°

FIG. 9: Graphic representation according to Spencer's Method of the SLOPE/W results describing the deterministic post-liquefaction conditions at Fruitgrowers Dam.

Tables 5 and 6 respectively summarize the results from the SLOPE/W analyses and the analyses run with PES. Figure 10 shows a direct comparison of the results from the two programs for both lower and higher v.

The results showed in Figure 10 outlines fundamental differences between the two programs. A detailed effort has been made during this study to comprehend the differences among the two programs, but while for the program PES a full version of the program's code is available, for the program SLOPE/W the author of this research has to solely rely upon the program manual, published by Geostudio, which does not provide detailed information on the program code.

Table 5. Results from the Fruitgrowers probabilistic analyses run with the program SLOPE/W.

(θ) m	Low v $p_f\%$	High v $p_f\%$
1.22	3.8	20.12
3.05	12.53	34.23
4.57	19.37	39.02
6.09	23.37	43.48
7.62	26.41	45.48
9.14	28.47	45.95
10.67	28.21	46.39
12.19	28.69	46.4
15.24	29.14	46.35
152.4	29.28	46.72

Table 6. Results from the Fruitgrowers probabilistic analyses run with the program PES.

(θ) m	Low v $p_f\%$	High v $p_f\%$
1.22	94.7	98.6
7.62	72.9	95.8
18.29	70.3	89
30.48	66.7	82.7
60.96	66.4	78.6
91.44	65.9	77.9
152.40	67.3	73.5

The p_f trend shown in Figure 10, corroborated by the trend results between program PES and the program SLOPE/W compared in the probabilistic validation presented in deWolfe (2010). The results presented in Figure 10 confirm that the probability of failure computed by SLOPE/W is unconservative with respect to the probability of failure estimated by program PES.

Figure 10 shows that for high values of spatial correlation the p_f results from both programs will show very little variation which is expected because high values of spatial correlation correspond to a virtually homogeneous soil material at each simulation. Lower values of spatial correlation instead emphasize a very different trend between the two programs.

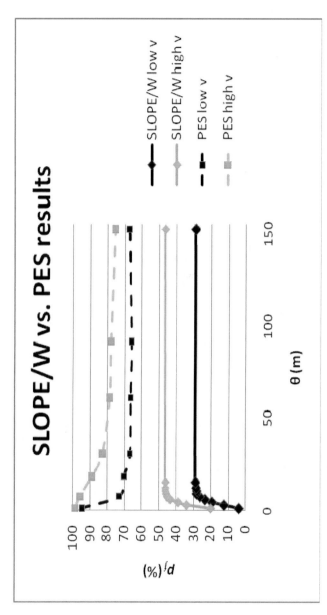

FIG. 10: Comparison of the results from programs PES and SLOPE/W for both lower and higher *v*. PES results are based on the deterministic F.S of 1.09 and the SLOPE/W results on the deterministic F.S of 1.06 To gain a better prospective on the comparison between the element size and the spatial correlation length in this model, it is important to remember that a single square element size is equal to 0.91 meter (3 feet), and the total dimensions of the problem are approximately, 21.95 meters (72 feet) in height and 146.3 meters (480 feet) in length.

The trend showed in Figure 10, by SLOPE/W results, associates lower p_f to a highly spatially variable soil (low spatial correlation) and a higher p_f with a more homogeneous soil (high spatial correlation). In the other hand, program PES show results which associate higher p_f with more variable soils and lower p_f to a more homogeneous soil. As mentioned in program SLOPE/W manual, the program does not apply any reduction to the standard deviation or the mean values of a random property unless the length between two sections, ΔZ, is equal to or greater than the scale of fluctuation or spatial variation length.

In the specific case of the model representing Fruitgrowers Dam the average distance between two slices is approximately 1.22 meters and therefore no reduction was ever applied to the standard deviation or the mean values of a random property through all analyses. In general in the case of a deterministic FS>1 a random field characterized by a reduced mean and variance values will lead to higher probability of failure, and that could explain why the SLOPE/W results are consistently unconservative with respect to the results computed by program PES. Instability in the results produced by program PES can be observed when the spatial correlation length value is equal to or smaller than the element size. In general, cases where the element size is greater than the spatial correlation length do not represents a very meaningful model, when instead, if many elements are able to define the variability inside the spatial correlation length, this can be considerate a representative model.

Even for the cases when this may apply, one unstable result certainly cannot in anyway change the overall interpretation of the analysis results trend.

Without a doubt it is quite difficult to determine the correct value of a soil variability and this parameter represents a key component of this probabilistic analysis. Only expert engineering judgment supported by exploration can truly lead to the understanding of what that meaningful range of soil variability is for a specific material. The results computed by the program PES and shown in Figure 10 clearly emphasizes that not accounting properly for soil variability will lead to unconservative results of p_f or non-convergence and underestimate the probability of slope instability. It needs to be remembered that the high probability of failure computed by program PES associated with Fruitgrowers dam is strictly dependent on the liquefaction of a continuous layer approximately 1.5 to 2 times the height of the embankment. Even though the presence of potentially liquefiable material has been corroborated by field testing in the area, the absolute continuity of the potentially liquefiable layer still remains uncertain. Furthermore, based on the blow counts values describing the strength of the weathered shale characterizing the potentially liquefiable layer, liquefaction can occur only for an event associated with a high seismic return period, such as the 50,000-year return period characterized by an acceleration value of 0.27g. The probability of such event occurring in this area is highly unlikely. For further information on the seismicity associated with Fruitgrowers Dam the reader is referred to the Bureau of Reclamation seismic study conducted in 2004 (Bureau of reclamation 2004).

CONCLUDING REMARKS

Program PES provides a repeatable methodology able to improve the confidence associated with the computation of probability of slope instability, which is a key component of risk assessment for an engineering structure.

The probabilistic approach used in program PES applies a combination of the random field technique and the finite element method.

At the core of the RFEM approach is the capability of accounting for spatially random shear strength parameters and spatial correlation. This methodology combines a non-linear elasto-plastic finite element analysis with random field theory generated using the Local Average Subdivision Method (Griffiths and Fenton, 2004). More specifically the spatially variable soil properties are correlated through the parameter spatial correlation length or scale of fluctuation (θ), which indicates the distance within which the values of a property show a relatively strong correlation, and the parameter correlation coefficient (ρ). The main advantage of the RFEM over traditional probabilistic slope stability techniques is that RFEM enables slope failure to develop naturally by "seeking out" the most critical mechanism.

The methodology utilized in program PES is compared against the probabilistic approach proposed by the program SLOPE/W version 7.14, and demonstrates its potential for predicting probability of failure in a non-homogeneous soil structure characterized by phreatic conditions and a possible liquefiable layer. While the results computed from the deterministic analyses using programs PES and SLOPE/W show a very close agreement, the results from the probabilistic analyses from the two programs are generally in disagreement, and the SLOPE/W results consistently show lower values of p_f than obtained using program PES.

In the author's opinion the difference in p_f computed by the two programs can be explained by the following three observations:

1. Both programs PES and SLOPE/W produce results of deterministic FS, pf, mean and standard deviation of FS, but it is important to remember that, for both probabilistic and deterministic analyses, program SLOPE/W represents a 1D model of the soil property correlations along the potential failure surface, while PES characterizes the soil property correlations using a 2D model. In the probabilistic approach, the program PES investigates the soil variability through the spatial correlation length over the entire foundation and embankment zones while SLOPE/W investigates the soil variability only along the line characterizing the critical slip surface.

2. Another major difference between the two programs is that SLOPE/W will perform the probabilistic analysis on a failure surface found using traditional slope stability methods (Jambu, Spencer, Bishop etc.) that require a subdivision of the slope into columns, while the program PES based on a strength reduction allows the modeled slope to fail naturally by "seeking out" the path of least resistance of each Monte-Carlo simulations. In the author's opinion, the number of columns initially selected by the user in program SLOPE/W not only influences the precision of the deterministic FS, but also influences the computation of the probability of failure.

3. Another component that may lead to the low values of probability by SLOPE/W, especially at lower values of the spatial correlation length (θ), is the difference in the way local averaging is implemented in the two programs.

The establishment of a robust methodology provided by this research will not only allow testing of the stability conditions of dams during modification phases, but will also help estimate the probability of failure in cases involving post-earthquake liquefaction. Although in the current study interest was concentrated on a classical two-sided embankment geometry, the methodology can be applied to a wide range of geotechnical engineering problems, taking into account the soil spatial variability and its capability of "seeking out" the critical failure surface without assigning a pre-defined failure surface geometry.

The current work has proven that not accounting for spatial variability can lead to unconservative results with respect to more classical approaches computing probability of failure in geotechnical problems.

REFERENCES

Borja A.W., Lee S. R., and Seed R.B. (1989) "Numerical simulation and Excavation in Elast-plastic soils". *Int J Numer Anal Methods Geomech*, 13930;231-249.

Bureau of Reclamation (BOR), November 2003. "Seismic hazard Assessment, Fruitgrowers Dam, Grand Mesa Project, Colorado" Technical Memorandum No. D-8330-2003-29, Bureau of Reclamation, *Internal publication*, Denver, Colorado.

Bureau of Reclamation (BOR), August 2004. "Fruitgrowers Dam Issue Evaluation: Evaluation of Liquefaction and Post Earthquake Stability" Fruitgrowers Dam Project, Colorado, Upper Colorado Region.Technical Memorandum No.FW-8312-3 Bureau of Reclamation, *Internal publication*, Denver, Colorado.

deWolfe, G. F. (2010). " Probabilistic and Deterministic Slope Stability Analysis by Finite Element. Unpublished dissertation. Colorado School of Mines, department of Engineering.

Fenton, G.A. and Vanmarcke, E. H. (1990). "Simulation of random fields via local average subdivision." *J Eng Mech, ASCE*, Vol. 116, No. 8, pp. 1733-1749.

Fenton, G.A. and Griffiths, D.V. (2008). "Probabilistic Methods in geotechnical Engineering" *Springer Wien* NewYork.

Geostudio Group, 2007. "SLOPE/W 2007 software for Slope Stability Analysis" *Geo-SLOPE International*,

Griffiths, D.V., and Fenton, G.A. (1993). "Seepage beneath water retaining structures founded on spatially random soil", *Géotechnique*, 43(6), pp. 77-587.

Griffiths, D.V. and Lane, P.A., (1999) "Slope stability analysis by finite elements". *Géotechnique*, vol. 49, no. 3, pp. 387-403.

Griffiths, D.V. and Fenton, G.A., (2004). "Probabilistic slope stability analysis by finite elements." *J Geotechnical and Geoenvironmental Engrg.*, vol.130, no.5, pp.507-518.

Lambe, T. W., Whitman R. V., (1969). "Soil Mechanics" *Wiley, New York*, pp. 553.

Matsui, T. and San, K-C., (1992). "Finite element slope stability analysis by shear strength reduction technique". *Soils and Foundations*, vol. 32, no. 1, pp. 59-70.

Phoon, K.-K., and Kulhawy, F.H., (1999). "Characterization of geotechnical variability". *Canadian Geotechnical Journal*, 36, pp.612–624.

Smith I. M. and Griffiths D.V., (2004). "Programming the Finite Element Method, 4th edn", *Wiley & Sons, West Sussex, England.*

Vanmarcke E.H., (1983). "Random Fields: Analysis and Synthesis". *MIT Press*, Cambridge, Mass.

Excavation Support and Micropile Underpinning in Vail, Colorado

Michael Shane Robison[1], A.M. ASCE

[1] Construction Manager, Schnabel Foundation Company, 2950 South Jamaica Court, Suite 107, Aurora, Colorado 80014; robison@schnabel.com

ABSTRACT: This case history describes the design and installation of excavation support and micropile underpinning for a building expansion and renovation project in Vail, Colorado. Local regulations required the building be preserved, thus leading to a new structure that was built both above and below the existing structure. A combination of soil nails and micropiles were needed to support the variety of earth and structural loading conditions that this project created.

Installing the micropiles proved to be more difficult than originally anticipated. Low overhead clearances and the tight corners made it extremely difficult to maneuver to the planned drilling locations. Additionally, there were old foundations and footings that were not discovered until the first attempt at drilling which made our initial planned drilling method impossible.

In order to adapt to the changed conditions, the method of drilling the micropiles had to be modified a couple of times. With each new method that attempted, more was learned about what worked the best until a system was identified that enabled the installation of the micropiles at the planned locations.

INTRODUCTION

Working in Vail, Colorado can provide many interesting obstacles. On this particular project, the owner wanted to add three additional stories to the existing structure while at the same time expanding the adjacent parking garage to extend below the existing foundation.

Micropiles were drilled through the existing footings to accommodate the additional load from the three additional levels. Most of the foundations were located in the existing lower level-parking garage. The parking garage had approximately eleven and a half feet of overhead clearance.

In addition to the original building, the site is surrounded by a variety of existing structures and improvements. The new basement level extended between three and eight feet below the existing foundation level of most adjacent structures. Temporary and permanent soil nail walls and soldier beams and lagging walls were installed to provide lateral earth support and minimize movement of these structures. The most critical design case included micropiles through the footings and soil nails in between the micropiles to accommodate the planned garage excavation below the structure.

SUBSURFACE INFORMATION

The subsurface conditions were typical of those encountered in mountain work in Vail, Colorado. They consisted of sands and gravels with scattered cobbles and boulders. Unfortunately there were no borings done inside the garage itself. There were some borings done on the outside of the existing garage and these were used to anticipate what types of material would be encountered during the drilling operation. The geotechnical report did not indicate the presence of any groundwater within the scope of the drilling.

MICROPILES

Micropiles are small diameter deep foundation piles. The equipment that is used to install them is fairly small which makes them ideal for situations that offer low overhead clearance and tight working conditions.

The micropiles were typically spaced 1.5 m (5ft.) apart. The loads on the micropiles ranged from 667 KN (150 Kips) to 3336 KN (750 Kips). The drilling depth of the micropiles varied from 4.3 m (14 ft.) to 10 m (33 ft.)

On this job, three different drilling methods were utilized and two different types of reinforcing bar were used. The first installation attempt used an air track drill with hollow core bar. The second installation attempt utilized a hydraulic drill with a down the hole hammer. The third installation attempt used a hydraulic drill with casing and a down the hole hammer.

AIR TRACK DRILL WITH HOLLOW CORE BAR

A low overhead track drill was used in the first attempt at drilling the micropiles in the existing parking garage. The micropiles used T76 hollow core bars as both drill rods during installation and as the final steel reinforcement. These bars have an outside diameter of 7.62 cm (3 in.) and an inside diameter of 5.08 cm (2 in.). A 15.24 cm (6 in.) diameter drill bit was used. During installation the air track's percussion rotary head simultaneously impacts and rotates the hollow core bars as neat cement grout is pumped through the center of the bar. The grout flushes cuttings from the bit face and returns to the top of the drill hole around the annulus of the hollow core bar.

The first three holes that were drilled only penetrated about twenty feet into the ground. Later on during the course of the project it was discovered that the hollow core bars were being drilled directly over an existing footing.

The drilling method described above was not successful because of the unknown obstructions. The air track could not provide enough impact energy to drill the T76 hollow core bars through the existing footings.

FIG: 1 Air Track Drilling Using Hollow Core Bar

After experiencing these delays, the production was well below what was estimated. In order for the contractor to be able to add the additional stories on to the existing structure all of the planned micropiles needed to be installed. The installation production of the micropiles needed to be increased to meet the project schedule. An alternate method of drilling would be required.

HUTTE 202 DRILL WITH DOWN HOLE HAMMER

A Hutte hydraulic rotary drill was brought in to replace the air track drill. Instead of hammering the steel into the ground, a down hole hammer was used and the original T76 hollow core bars were replaced with #18 grade 75 solid bars. The bars were inserted in the holes after they were drilled.

A 15.24 cm (6 in.) down the hole hammer with a 15.24 cm (6 in.) bit was used to advance the hole. The drill rods used were API drill rods. Casing was not used with the first attempt with the down the hole hammer.

The geotechnical report stated the water table was well below the drilling area of the micropiles. However, the first hole encountered a significant amount of water that caused the drill holes to cave. It was apparent the holes could not be drilled without casing.

Additionally, stronger drill rod was needed to accommodate the powerful hydraulic rotary head of the Hutte.

FIG.2. Hydraulic drill rig

HUTTE 202 DRILL WITH DOWN THE HOLE HAMMER AND CASING

Due to the lack of success with the previous two micropile installation methods, the drill steel was changed to 6.03 cm (2 3/8 inch) API pipe to allow drilling both six and four inch holes with the advancement of casing. The 15.24 cm (6 in.) hammer was used to drill through the underlying retaining wall and existing footings. It was then replaced with a 10.16 cm (4 in.) hammer and 13.3 cm (5.24 in.) casing. Carbide-tipped cutting teeth on the casing crown were used to aid in advancing the casing.

Because of low overhead conditions, 1 m (3.3 ft.) lengths of drill rod and casing were used. Advancing the casing and the drill rod created some challenges. The drill rig that was being used only had a single head on it, making it difficult to add the casing and the drill rod at the same time. A shorter piece of casing was fabricated to facilitate adding both the steel and the casing. Once this drilling system was established, the production on the micropiles proceeded efficiently and timely.

PERMANENT SOIL NAILS AND SHOTCRETE

Upon completion of the micropiles, excavation could proceed below the existing foundation. Permanent soil nails and shotcrete were used to support the excavation. Soil nails and shotcrete is a top down method of shoring. The excavation proceeds in five foot lifts or smaller in order to keep the material from the previous lift from sloughing off. Soil nails were installed on every lift. Care had to be taken as they were installed to not drill through the micropiles that had been previously installed.

The walls were designed for permanent earth retention only. They were not designed for surcharge loading of the building that was carried by the micropiles.

On each lift after the soil nails were installed a permanent shotcrete facing was applied. The soil nails were hollow core bars that were drilled in place with grout. A top drive hammer is used to beat them into the ground while grout is being flushed through them. The shotcrete was 20.3 cm (8 in.) thick and was reinforced with welded wire mesh and #4 rebar. The shotcrete was then screeded for a flat finish. The permanent walls were incorporated into the new building.

TEMPORARY SHORING

In addition to the permanent support of the existing structure, excavation was planned below the foundations of most of the adjacent structures. Care had to be taken as the excavation proceeded to minimize settlement of these structures. Most of the surrounding buildings were supported on structural fill which typically had very little cohesion. In order to stabilize this material additional hollow core bars were inserted and extra grout was injected to solidify the mass of fill. Temporary soil nails and shotcrete were used to support these structures. Once again hollow core bars were used to provide the support for the shoring wall. A 10.16 cm (4 in.) shotcrete wall was installed for support of the lateral earth pressure. In some instances where it was not possible to put nails across the property line, temporary cantilever soldier beams and lagging were used. The temporary shoring held the cut open around the perimeter of the entire new parking garage while the new foundation was constructed and was designed for earth pressure only. The height of the temporary shoring varied form 1.83 m (6 ft.) to 7.62 m (25 ft.) Upon completion of the new structure above the existing grade, the building was ready to assume the load that the temporary shoring was holding and the area between the new walls and the temporary shoring was backfilled.

FIG.3. Temporary soil nail wall

MICROPILE TESTING

Testing of the micropiles within the congested parking structure proved to be a difficult task. There were not very many areas in which micropiles were configured to provide the space needed for a micropile test on a production pile.

Once a suitable location was established the test was set up. The testing performed on the piles was compression testing per ASTM D1143. In order to do the test three micropiles were needed that were far enough apart so as not to interfere with each other while the test was being conducted. A large whaler was placed over the three micropiles. The two outside piles were used for reaction piles while the middle pile was subjected to the compression load. A hydraulic jack was placed on the middle pile under the whaler. The jack pushed against the whaler and down on the micropile.

Dial gauges that read to the nearest .00254 cm (.001 in.) were placed on plates attached to the micropile to measure the deflection of the micropile. The micropile was then incrementally loaded from 25% of its design load to 200% of its design load. Because it was not possible to test all of the micropiles only certain ones that met the criteria mentioned above were tested. These micropiles had a design load of 2210 KN (497 Kips).

The test criteria for this project stated that the top of the micropile could not move more than .05 inches during the ten minute load hold at 133% of the design load. The movement on the production piles ranged from zero to .02 inches of movement.

FIG.4. Micropile compression test

FIG.5. Dial gauges for micropile test

CONCLUSIONS

Whenever working underground, there is always the possibility for unknowns that may be unquantifiable. In this instance it was very important to have a backup plan for the installation of the micropiles.

Realizing that your first try is your best guess but that it is probably going to have to be modified is a realistic way of approaching a job.

When the first attempt at installing the micropiles did not meet the production goals, immediately backup plans were used and then modified to the end result that allowed for the completion of the job. It is also important to realize that even though the first attempt did not go as planned the method of drilling that was used to complete the job was a combination of everything that was tried leading up to that method.

ACKNOWLEDGMENTS

The author recognizes the support of both Schnabel field and office personnel whose knowledge, advice and experience were greatly appreciated.

The Use of Geophysical Methods to Detect Abandoned Mine Workings

James W. Niehoff[1], M. ASCE, P.E.

[1]Geotechnical Practice Leader, Golder Associates Incorporated, Lakewood, CO 80228; jniehoff@golder.com

ABSTRACT: Prior to the mid 1900s, significant coal mining was conducted in the Front Range of Colorado. Most of the mines extracted coal from relatively thin seams located at shallow depths using room and pillar mining techniques. Subsequent to mine abandonment, the remaining coal pillars and weak rocks above the "rooms" have tended to degrade and ultimately cause either gradual or abrupt subsidence of the overlying ground. Development in these undermined areas must consider the potential risk and magnitude of future subsidence. Mine subsidence studies in Colorado have traditionally employed widely spaced exploratory borings to detect the thickness and extent of remaining voids in mined areas. However, due to the limited area explored, such studies do not necessarily provide sufficient data for an accurate assessment of future mine subsidence. In recent years, geophysical testing has been used as a tool to initially screen undermined sites in an effort to detect the location of possible voids or disturbed geologic strata. Using geophysical data, borings may be located in areas underlain by anomalous conditions, rather than on an arbitrary grid pattern. Two methods that appear to show promise for use in this regard include Refraction Microtremor and multi-channel resistivity testing. This paper presents several project examples that demonstrate the use, effectiveness and limitations of these geophysical testing techniques within undermined sites north of metropolitan Denver.

INTRODUCTION

Beginning in the 1860s and continuing into the mid-20th Century, extensive underground coal mining was undertaken in the Front Range of Colorado. One of the most productive areas for coal mining was in Boulder and Weld Counties to the north and northwest of Denver. Here, a number of coal seams, typically 1.2 to 2.5 meters (4 to 8 feet) in thickness, are present at depths ranging from about 15 to 90 meters (50 to 300 feet) below the ground surface.

Coal was generally extracted employing the room and pillar technique. In this approach, shafts were excavated to the depth of the coal seams. From these, passageways were excavated. Individual rooms typically about 60 meters (200 feet) in length and about 5 to 5.5 meters (16 to 18 feet) in width were then excavated into the coal seams perpendicular to the direction of the passageways. Pillars of coal about 5 to 5.5 meters (16 to 18 feet) in width were left in place between the rooms to support the sedimentary claystone and sandstone bedrock comprising the roof of the mines.

Once primary mining was completed, the pillars were often "pulled" to increase the overall area of coal mined from about 50 percent to 70 percent or greater. Most of the mines in the Boulder-Weld County coal field were closed and abandoned in the late 1800s and early to mid-1900s.

Over time, the abandoned mines typically experience partial or total collapse as a result of the decay of support timbers, softening of the rock comprising the roof and floors, and gradual spalling of coal pillars. Initially, as the exposed rock is subjected to wetting from groundwater, it softens and large pieces fall from the roof to the floor. The resulting pile of rock fragments is loose with a high percentage of voids. Spalling of roof rock tends to continue upward in this manner a vertical distance of 8 to 10 times the thickness of the coal seam, when a point of some equilibrium is reached.

Depending upon the thickness of the coal seams, the percentage of coal mined, the depth of the mine below the ground surface, and the thickness and consistency of the bedrock above the mine, there can be minor to significant surface manifestations associated with mine collapse. The most dramatic features resulting from the collapse of abandoned mines are sinkholes. These typically form in areas where the mined intervals are shallow, where the rock above the mined interval is less than 8 to 10 times the coal seam thickness, and where surface soils are granular and prone to erosion. Sinkholes can also form where shafts are improperly plugged following abandonment. Figure 1 below shows a sinkhole about 12 meters (40 feet) in diameter and 9 meters (30 feet) deep that formed over a shallow abandoned mine in Erie, Colorado.

FIG. 1. Large mine-induced sinkhole in Erie, Colorado

More commonly, collapse of mined rooms and passageways results in more gradual ground subsidence. Where mines are at great depth and coal seams are relatively thin, surface subsidence may be insignificant or distributed over a wide area. However, where coal is mined from shallow depths, subsidence may be evidenced at the surface as relatively narrow troughs or closed depressions that mirror the shape and configuration of underlying mined rooms.

Mine collapse and associated ground subsidence can have significant consequences to overlying residential and commercial structures, as well as roadways and utilities. As a consequence, development in and near undermined areas requires significant study to define the extent and condition of abandoned mines to ascertain risks of future ground movements.

TRADITIONAL ASSESSMENT METHODS

The assessment of the potential for mine subsidence begins with a review of available mine maps. Unfortunately, such maps depict shafts, passageways and other features that are not necessarily located accurately relative to current ground coordinate systems. Further, such maps may depict the extent of the mines several years prior to their abandonment. Consequently, they are generally considered suitable for assessing the general location, size and depth of an abandoned mine, but are often not considered accurate on a local, site specific basis.

Review and interpretation of aerial photographs (particularly stereo pairs) can provide indications of surface subsidence patterns indicative of mine collapse. These are most useful for relatively shallow mines, where ground deformation is significant. Surface vegetation and land disturbance activities, such as those associated with farming, can mask minor surface subsidence features.

Accepted local practice for the assessment of subsidence risk involves drilling widely spaced boreholes within and adjacent to areas that are documented as having been undermined. Boreholes are logged both visually and with downhole geophysical equipment to define the depth to and thickness of coal seams. Borehole calipers are used to detect variations in borehole diameter and the presence and vertical extent of open voids. While the information obtained at specific borehole locations can be accurate using these procedures, it is equally likely that boreholes will encounter pillars as open seams with a typical mine extraction rate of 50 percent.

GEOPHYSICAL SCREENING

Geophysical testing as described in this paper is not intended to replace the use of borings, but to serve as a screening tool to increase the likelihood that borings will penetrate abandoned passageways and mined rooms. Geophysical tests can also provide data relative to the extent, both vertically and laterally, of rock collapse above mined seams. The two methods that have been used with some success in this area include Refraction Microtremor testing and Resistivity testing.

Refraction Microtremor (ReMi)

Refraction Microtremor (ReMi) testing involves the measurement of surface waves along a straight line array with 24 or more 10-Hz geophones spaced at 3 to 10 meter (10 to 30 foot) horizontal intervals. Rather than using a specific vibration source as employed in conventional seismic refraction tests, noise from nearby vehicle traffic and other background sources is employed. The acquired data is analyzed by means of the SeisOpt ReMi™ computer program developed by Optim, Inc. The program employs a wavefield transformation data processing technique and an interactive Rayleigh-wave dispersion modeling tool, which exploits the most effective aspects of the spectral analysis of surface waves.   The slowness-frequency wavefield transformation is particularly effective in allowing the accurate selection of Rayleigh-wave phase-velocity dispersion curves despite the presence of waves propagating across the linear array at high apparent velocities, higher-mode Rayleigh waves, body waves, air waves, and incoherent noise. A series of 10 to 20 records, each about 20 seconds in length, allows for the stacking of data, and better definition of the shear wave velocity profile across the array. The ReMi method differs from many other seismic tests in that it allows the detection of low velocity materials beneath high velocity strata. A typical ReMi profile generated from data obtained in a mined area is presented in the figure below.

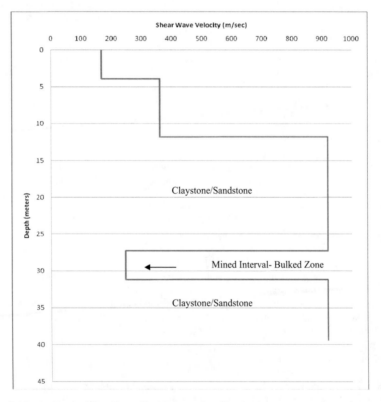

FIG. 2. Typical ReMi profile in an undermined area – note low shear wave velocity zone between 25 and 28 meters (82 to 95 feet).

By selecting and evaluating sets of data for groups of adjacent geophones along the array, a two dimensional cross-section of the subsurface profile may be developed. This profile can generally define the vertical and lateral extent of larger low velocity zones in the subsurface, indicating the potential presence of passageways and mined rooms.

Electrical Resistivity

Electrical resistivity testing introduces a current into the ground and measures the resistance of the ground to the electrical field. The size of the field and the depth of penetration is directly related to the spacing of the electrodes along the array. In areas underlain by relatively level strata, distinct patterns of electrical resistivity can be detected and interpreted. Voids or very loose materials within these strata are very resistant to the flow of electrical current when such materials are dry. Conversely, voids and low density zones below the water table are highly conductive. Through interpretation of electrical resistivity arrays, two-dimensional profiles may be developed that may reveal anomalous areas of low or high resistivity within otherwise uniform strata. While this technique has found significant use only recently in mine detection, it has been used for many years in the detection of other types of subsurface anomalies and voids such as those present in karstic geologies.

Limitations

Neither of the geophysical testing techniques noted in the paragraphs above yield data that allow for the development of unique solutions. That is, many subsurface profiles may fit the data equally well. Consequently, some information relative to subsurface conditions and the depth to likely mined seams needs to be considered to provide for a reasonable interpretation of the data. It should also be noted that these techniques lose accuracy with depth. In general, accurate data interpretation below a depth of about 38 to 46 meters (125 to 150 feet) is not feasible. The geophysical techniques noted herein are thus best suited to relatively shallow mines, where coal seams are within about 30 to 38 meters (100 to 125 feet) of the ground surface.

CASE HISTORIES

Two case histories are presented herein to demonstrate the usefulness of geophysical methods as a screening tool for mine subsidence studies. The first of these was a study for a proposed shopping center in Erie, Colorado, while the second was conducted for a proposed hospital complex near Frederick, Colorado.

Shopping Center Site

An approximately 8 hectare (20-acre) site in Erie, Colorado was selected by a major retail developer for a potential shopping center. Initial studies suggested that significant portions of the site were underlain by a coal mine. While mine maps were available, it was clear that they were incomplete as many of the passageways and rooms were not depicted with clear end points. Research suggested that mining activity had taken place from the latter part of the 19th Century through the early part of the 20th Century and had removed coal from one or more seams about 26 to 36 meters (85 to 120 feet) below the ground surface. The maps also suggested that pillars had been pulled from substantial areas of the mine prior to closure. An abandoned and backfilled mine shaft was evidenced by the presence of an earthen mound near the southwestern corner of the site. An overlay of the mine and proposed development is shown on Figure 3.

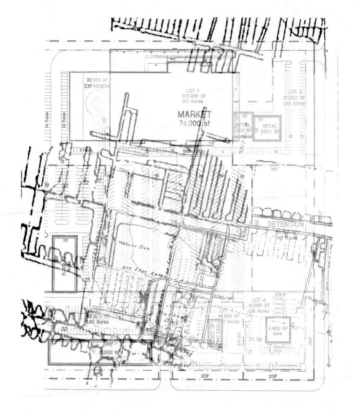

FIG. 3. Mine map overlay on shopping center development plan.

To provide for an overall assessment of the site, a total of 8 ReMi lines 275 meters (900 feet) in length were conducted on site in a north-south grid pattern with lines approximately 46 meters (150 feet) apart. An additional 91-meter (300-foot) long line was conducted in a perpendicular direction through the site of a proposed strip center near the southwestern corner. Each ReMi line employed 10-Hz geophones spaced 4.6

meters (15 feet) apart. Data were collected employing a DAQ Link II 24-channel seismic acquisition unit and the VScope computer program distributed by Seismic Source, Inc. Data were reduced and 2-dimensional profiles were generated. A typical profile is presented in Figure 4.

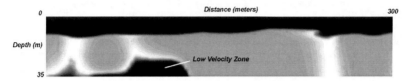

FIG. 4. Typical ReMi cross-section at Erie shopping center site.

The sections generally revealed a zone of low to moderate shear wave velocity material to a depth of about 7.5 to 9 meters (25 to 30 feet) below grade, representative of overburden soil. This was underlain by a moderately high shear wave velocity material, interpreted to be sedimentary bedrock. In some areas, lower velocity zones were indicated within the bedrock zone. These lower velocity zones were generally interpreted to extend to depths of about 33.5 to 36.5 meters (110 to 120 feet) below grade. Areas where very low velocity materials were encountered were identified for further exploration by soil test borings. Borings generally confirmed the thickness of the overburden soils corresponding to the near surface low shear wave velocity material and the presence of claystone/sandstone bedrock below. Within the bedrock zone in locations where only moderate shear wave velocities were interpreted, samples suggested some delamination/stress relief of bedrock strata as noted in Figure 5.

FIG. 5. Split spoon sample of claystone/sandstone bedrock suggesting delamination and stress relief

At the depth indicated to have very low shear wave velocity material, the drilling augers encountered very little resistance, suggesting a loosely filled void. In one location, an additional boring was drilled about 3 meters (10 feet) away and encountered an intact coal pillar.

Within this site, the ReMi approach was considered very useful in confirming the depth of the mined coal seams, and evaluating the lateral extent of the mine within the site. With the data obtained by the ReMi method, the locations of exploratory boreholes were selected to characterize the subsurface profile in mined and undisturbed areas. This allowed for a more efficient evaluation of the potential for and extent of mine related subsidence within the site.

Medical Center Site

A large parcel adjacent to Interstate 25 in Frederick, Colorado was considered by a large hospital corporation for the development of a new medical center. This site was mapped as being underlain by a coal mine at depths of 30.5 to 38 meters (100 to 125 feet) below grade. Preliminary studies were conducted using widely spaced conventional soil borings with highly variable results. Many borings did not encounter mine workings to depths of 39.5 meters (130 feet) or deeper. Others found rubble zones above the depth of mined coal seams. A few found small open voids. To provide for a better overall assessment of the site, both ReMi and resistivity tests were conducted within this site. For this study, a SuperSting R1/IP instrument console made by Advanced Geosciences, Inc. was employed for the resistivity testing. For each test, dipole-dipole arrays were employed with 28 electrodes spaced at 6 meters on center. Data were reduced and interpreted by means of AGI EarthImager 2D software.

The geophysical techniques revealed zones of low shear wave velocity and low resistivity beginning at depths of approximately 25 meters (80 feet) below grade. A typical resistivity profile is presented in Figure 6.

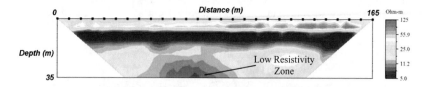

FIG. 6. Resistivity profile indicating zone of low resistivity indicative of a mined seam and bulked rock above.

Soil test borings were subsequently drilled in areas revealed to have anomalous subsurface conditions. In most instances, these borings confirmed the presence of mined seams and overlying bulked rock zones consistent with the geophysical data. In a few instances, mine workings were not encountered in borings in areas where geophysical testing suggested the presence of voids or loose materials. This lack of correlation was attributed to the averaging effects of geophysical interpretation, particularly at significant depth. That is, the geophysical data reflect conditions within a relatively wide zone on either side of the ReMi or resistivity array. In spite of this shortcoming, the geophysical testing program was useful in confirming the approximate lateral extent of the mine, the depth to the mined interval, and the thickness of disturbed and relatively intact bedrock above the mined seam. These are all critical characteristics in the assessment of potential mine subsidence.

CONCLUSIONS

In summary, surface geophysics can play a significant role in screening sites for prior mining activity. The ReMi and resistivity techniques described herein are particularly well suited to sites where mining was carried out within 30.5 to 38 meters (100 to 125 feet) of the ground surface. These methods can provide useful data relative to the depth to the mined interval, the lateral extent of mining, and the thickness and continuity of bedrock above the mine. While geophysical testing will not replace deep borings in the assessment of mine subsidence risk, it can serve as a supplemental source of useful data to characterize subsurface conditions and to allow for the efficient selection of boring locations.

ACKNOWLEDGMENTS

The author appreciates the assistance of Matt Satterfield and Kyle Duitsman of PSI, who generously furnished much of the case history data used in this paper.

REFERENCES

Anderson, N., Croxton, N., Hoover, R., and Sirles, P. (2008). "Geophysical methods commonly employed for geotechnical site characterization." Transportation Research Circular E-C130, Transportation Research Board, Washington, DC.

Ishankuliev, M., (2007). "Resistivity imaging of abandoned minelands at Huntley Hollow, Hocking County, Ohio." A thesis presented to the faculty of the College of Arts and Sciences of Ohio University in partial fulfillment of the requirements for the degree Master of Science.

Johnson, W. (2003). "Applications of the electrical resistivity method for detection of underground mine workings." Geophysical Technologies for Detecting Underground Coal Mine Voids, Lexington, KY.

Louie, J. (2001). "Faster, better: shear wave velocity to 100 meters depth from refraction microtremor arrays." Bulletin of the Seismological Society of America.

Putnam, N., Nasseri-Moghaddam, A., Kovin, O., and Anderson, N., (2008) "Preliminary analysis using surface wave methods to detect shallow manmade tunnels." Proceedings, 21st Symposium on the Application of Geophysics to Engineering and Environmental Problems, Philadelphia, PA.

Sherman, G., and Partington, B. (2006). "Abandoned mine subsidence prediction using British National Coal Board methods, Denver, Colorado." (IAEG2006), Paper number 640.

Zhou, W., Beck, B.F. and Stephenson, J.B. (2000). "Reliability of dipole-dipole electrical resistivity tomography defining depth to bedrock in covered karst terraines." Environmental Geology (39): 760-766.

Author Index

Subject Index

Page number refers to the first page of paper